Composition Research

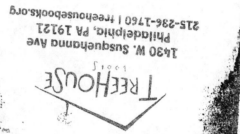

Composition Research

Empirical Designs

JANICE M. LAUER

J. WILLIAM ASHER

New York   Oxford
OXFORD UNIVERSITY PRESS
1988

Oxford University Press

Oxford   New York   Toronto
Delhi   Bombay   Calcutta   Madras   Karachi
Petaling Jaya   Singapore   Hong Kong   Tokyo
Nairobi   Dar es Salaam   Cape Town
Melbourne   Auckland

and associated companies in
Beirut   Berlin   Ibadan   Nicosia

Copyright © 1988 by Janice M. Lauer and J. William Asher

Published by Oxford University Press, Inc.
200 Madison Avenue, New York, New York 10016

Oxford is a registered trademark of Oxford University Press

Library of Congress Cataloging-in-Publication Data
Lauer, Janice M.
   Composition research: empirical designs / Janice M. Lauer and J.
William Asher.
     p. cm.
   Bibliography: p.
   Includes index.
   ISBN 0-19-504171-2. ISBN 0-19-504172-0 (pbk.)
   1. English language—Rhetoric-Research.   2. English language—
Composition and exercises—Research.   3. English language—
Rhetoric—Study and teaching.   4. English language—Composition and
exercises—Study and teaching.   I. Asher, J. William.   II. Title.
PE1404.L34 1988
808'.042'072—dc19   87-23054
                                                              CIP

Printing (last digit): 9 8 7 6 5 4 3 2 1
Printed in the United States of America
on acid-free paper

Foreword

Although people have been writing for centuries, and schools and universities have been teaching the arts and skills of writing for a long time, writing and the teaching of writing have only recently been the objects of scrutiny by empirical researchers. One reason for this fact, of course, is that empirical research is a relatively young field, but even despite its youth, research on the teaching of writing and the performance of writers has blossomed only within the past two decades. This sudden growth follows the growth of writing instruction and the evaluation of writing performance into vast enterprises. Writing texts and programs form a staple of many publishers; writing assessments are given to some 10 million students annually; departments of English at colleges and universities and—to a lesser extent—secondary schools, depend on composition programs for their health, if not their existence.

At the same time, scholars in various fields, notably anthropology and psychology, but also history and sociology, have come to develop an interest in writing and writing instruction, in part because writing is one of the less well-explored aspects of human behavior and at the same time one of the most central, particularly in complex industrialized societies. In these societies, a good writer—by whatever standard—is fairly sure to be successful in the affairs of society. Good writing tends to guarantee entry into the elite group, because writing performance is one of the foremost manifestations of literacy. And in a society that seeks to prepare a large segment of the population for elite status through its educational institutions, writing instruction bears great import. As an area of study in education, then, writing has tended to be studied from the broad social science perspective that has marked educational studies in the United States during this century. Educational research has been dominated by the experimental approach of psychology and, to a lesser extent, the approaches of sociology and anthropology.

Teachers of writing, however, and administrators of writing programs (particularly in higher education) have usually been trained in various types of humanistic study that have traditions and methodologies of research different from those of the social sciences. Yet they are asked to provide information about their activities and programs and about their students' performance and abilities to audiences that are used to information being presented in the traditions of social science research.

Humanists know that the writer sitting with a pen or at the keyboard is an intensely lonely individual; on the other hand, educational administrators are interested less in the individual than in the group. After all, writers in schools are in groups. Writing teachers want to reach the individual, to be sure, but given the numbers they have to cope with, they are compelled to consider the class. The whole educational enterprise, therefore, suggests that the traditions of humanistic research must be supplemented by social science research when one comes to examine the teaching of writing, writers, and writing in an educational context. That to me is the primary reason for this book's existence; it serves as a guide to new cultures.

Earlier I wrote of the dominance in educational research by the psychologist and the anthropologist; those two disciplines are emphasized in this book. As disciplines they must be defined, for their characteristics differ markedly. In an article on cross-cultural psychology, Herbert W. Barry III has written that "Psychology constitutes the scientific study of individual behavior in a social context" (Barry, 1979, p. 208) and that "Anthropology constitutes the descriptive study of all aspects of an exotic culture" (p. 209). He summarizes their differences in the following table (p. 208)—to which I have added the first row:

	Psychology	Anthropology
Tradition	Positivist–scientific	Aristotelian
Topic	Individual behavior	Exotic culture
Technique	Quantitative analysis	Fieldwork
Principal contribution	Statistical inference	Comprehensive observation
Principal limitation	Single variable and culture	Lack of analysis or comparisons

Each discipline complements the other: as the authors suggest, case studies and ethnographies can present comprehensive portraits of individuals or groups situated within their "exotic" culture; descriptive and experimental studies of a psychological nature can present portraits of behavior within a theoretical framework that speaks of "normal" or "expected" behavior. The first kinds of study are limited in that they cannot tell us whether the exotic culture is typical; the second kinds cannot tell us whether the normal is culturally determined. Together, they can provide writing teachers and planners of programs concerning writing with much useful information as to how to cope with the paradox of the lonely crowd of writing students.

For the humanist teacher of writing, the anthropological perspective appears most familiar: it resembles a narrative; it uses language without tables and statistics; and it appears to resemble the typical literary essay that also describes an exotic culture. The apparent familiarity of the case study (that may derive from Freudian psychology more than from anthropology) or the ethnography in part explains why so many English teachers turned researchers find the approach congenial. As the authors observe, however, the rigor of the ethnographer is quite different from the speculative turn that marks the literary critic. The number of ethnographic studies of writing that have the rigor of the anthropological studies is quite limited. Even the two examples discussed in the book lack the thoroughness of description that marks a work like Shirley Brice Heath's *Ways with Words*.

When educational researchers turn to the world of the psychologist and the experimental or quasi-experimental study, they often fall short not in their statistical design, but in their handling of the treatment. Such is also true of the descriptive studies, to be sure, but the experimental researcher in education most frequently fails in the reporting of the treatment. In psychology, as in the physical and biological sciences, the description of the treatment is, perhaps, the most crucial aspect of the study, for the experiment must be replicable. Most psychological studies give detailed descriptions of what went on in the treatment and also describe quite fully the nature of the control group. Unfortunately few if any of the experimental studies in the teaching of writing can be replicated, because the records of what goes on in the classroom are inadequate or lost. One problem, of course, is that the length of time of educational treatment is quite long, and certainly such record keeping would be tedious for the researcher and boring for the reader.

Another problem facing those who undertake descriptive statistical or experimental studies lies in the models that underlie the study. Statistical manipulations of data are used to test models of behavior, usually either relational models or causal models. One sets forth what appears to be a rational if not logical set of relationships among the phenomena that one is describing or manipulating, and the design of the study is such as to test the strength or even the existence of those relationships. Many of the empirical studies in writing do not appear to have a model that is tested, or they have a model that appears flawed.

This volume will help many of those in the field of writing who are about to embark on a research study as well as those who read research studies in order to make decisions about their own instruction or the program of instruction of which they have charge. It clarifies technical issues and it clearly shows the pitfalls that beset researchers no matter what sort of a study they are undertaking. To the extent that they can, the authors suggest which research method is appropriate to what sort of question, but often there are a number of ways by which that question can be approached.

I was recently asked to assist the staff at a large university who had been charged with examining a program in which the freshman composition course was being given in high schools and taught by high school teachers. The question they had was whether the high school course was equivalent to the college course. That is a broad question and it could be narrowed in a number of ways: are the behaviors of teachers and students similar? have the two groups of students learned the same things? is the written performance of the two groups of students the same upon completion of the course? is there a similar pattern of growth in the two groups of students? and so on. Each of these questions is an important one, but no two can be answered using the same research design. And in each of the questions there lie issues of definition and design. Whatever question lies at the heart of a research study dealing with this general issue, the person undertaking the research must turn the question into a research design with a model that can be tested. Having digested this volume, that person would be well equipped to undertake a study that has a reasonable chance of satisfying the demands of the questioners.

State University of New York at Albany   Alan C. Purves

References

Barry, H.W. III. (1979). Forecasts on the evolution of cross-cultural psychology and anthropology. In L.H. Eckensberger, W.J. Lonner, & Y.H. Poortinga (Eds.). *Cross-Cultural contributions to psychology* (pp. 207–18). Lisse: Swets and Zeitlinger.

Preface

Composition instructors, largely trained as humanists in departments of English, have responded to empirical research on writing in a variety of ways. At best, some have made extensive efforts to educate themselves in research design, statistics, and measurement in order to become researchers or discriminating readers of empirical studies. Others have responded in two extreme ways: either dismissing all empirical research or accepting its conclusions indiscriminately. We maintain that an adequate study of the complex domain of writing must be multidisciplinary, including empirical research. Improvements in instruction and advancements in knowledge about writing will stem from communication among composition theorists, writing instructors, and empirical researchers, who come to respect and value each other's efforts.

We concentrate in this book on one major dimension of empirical research—its designs—because they often become obstacles to understanding for the humanist. Although we introduce specific composition studies to help clarify technical concepts and procedures, we do not undertake to provide a comprehensive overview of the exploding number of studies being conducted on numerous aspects of writing today. We organize our bibliographies around research designs instead of around specific writing areas. Nor do we attempt to provide critical analyses of the constructs, variables, or underlying theories of writing that characterize specific composition studies. Such analyses, which constitute another important treatment of empirical research, have been done by a number of others recently including Beach and Bridwell (1984), Scardamalia and Bereiter (1986), Faigley et al. (1985), Hillocks (1986), and Moran and Lunsford (1984). (See references at the end of Chapter 1.)

In our effort to explain eight major research designs, one of our challenges has been terminology—trying to create bridges between the languages of the humanist and social scientist. Over the last four years, we have received helpful responses to our translation efforts from our colleagues and graduate students at Purdue. In particular, we want to thank the following who have read the manuscript at various stages and have provided valuable advice and help: Janet Atwill, Michael Carter, Kathryn Cochran, Jeanne Halpern, Anne Rosenthal, Patricia Sullivan, Irwin Weiser, and Leonara Woodman. We owe a special debt to Edward Vockell for his astute and detailed criticisms of the technical features of the manuscript. Our gratitude is also great to the authors of the studies whose research has made contributions to the field and offered us fine illustrations of the designs. We are grateful to the Literary Executor of the late Sir Ronald A. Fisher, F.R.S., to Dr. Frank Yates, F.R.S., and the Longman Group Ltd, London, for permission to reprint Tables III, IV, V, VI, and XXXIII from their book *Statistical Tables for Biological, Agricultural and Medical Research* (6th Edition 1974). We are also indebted to our editors at Oxford University Press for their support and advice.

Finally, we wish to thank our spouses, David Hutton and Kay Asher, for their patience, encouragement, and support throughout five years of work.

West Lafayette, Ind.   J.M.L.
July 1987   J.W.A.

Acknowledgments for Use
of Extended Passages

Bamberg, B. (1978). Composition instruction does make a difference: A comparison of the high school preparation of college freshmen in regular and remedial English classes. *RTE, 12,* 47–59. Copyright © 1978 by the National Council of Teachers of English. Reprinted by permission of the publisher.

CCCC Committee on Teaching and its Evaluation in Composition. (1982). Evaluating instruction in writing: Approaches and instruments. *CCC, 33,* 213–28. Copyright © 1982 by the National Council of Teachers of English. Reprinted by permission of the publisher.

Davis, B., Scriven, M., & Thomas, S. (1981). *The evaluation of composition instruction,* pp. 13–24; 67–148. Copyright © 1981 by Edgepress. Reprinted by permission of Edgepress.

Eblen, C. (1983). Writing across the curriculum: A survey of a university faculty's views and classroom practices. *RTE, 17,* 343–48. Copyright © 1983 by the National Council of Teachers of English. Reprinted by permission of the publisher.

Emig, J. (1971). *The composing processes of twelfth graders.* Copyright © 1971 by the National Council of Teachers of English. Reprinted by permission of the publisher.

Fox, R. (1980). Treatment of writing apprehension and its effects on composition. *RTE, 14,* 39–50. Copyright © 1980 by the National Council of Teachers of English. Reprinted by permission of the publisher.

Glass, G.V, McGaw, B., & Smith, M.L. *Meta-analysis in social research.* Copyright © 1981 by Sage Publications, Inc. Reprinted by permission of Sage Publications, Inc.

Halpern, J. Liggott, S. (1984). *Computers and Composing.* Copyright © 1984 by the Conference on College Composition and Communication. Reprinted by permission of the publisher.

Abbreviations in References

CCC	*College Composition and Communication*
DAI	*Dissertation Abstracts International*
RTE	*Research in the Teaching of English*

Contents

Composition Research

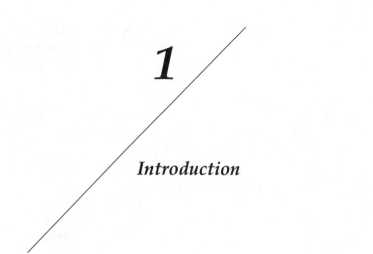

1

Introduction

Our purpose in this book is to explain eight empirical research designs available for composition inquiry. Our goal is to enable readers without prior empirical training to discriminate among types of research, to examine studies with useful criteria, and to select designs appropriate for their own situations. We neither assume statistical expertise in our readers nor provide extensive instruction in statistics and measurement, but do indicate those procedures and measures that are appropriate for each design. Our secondary audiences include those who already engage in empirical research on writing and administrators who evaluate such inquiry. Finally, we want to familiarize readers with a number of current composition studies using each design. Our bibliographies, which begin where Braddock, Lloyd-Jones, and Schoer conclude (1963), are not comprehensive but rather illustrative of the eight designs.

Modes of inquiry in composition research

Before discussing empirical designs, we think it important to point out that empirical research is only one of several types of research being conducted in composition studies. Other modes of inquiry include historical, linguistic, philosophical, and rhetorical. Each of these modes has offered insights into the nature of writing: its processes, development, facilitation, variations, and interactions with various contexts.

Vitanza (1987) contrasts traditional historiography with revisionary and sub/versive historiographies. The first two volumes of van Dijk's *Handbook of Discourse Analysis* (1985) offer a comprehensive treatment of one dimension of linguistic inquiry. Phelps (1983) analyzes the nature of philosophical inquiry in composition. Less well understood is rhetorical inquiry, one of the earliest and most common types used to develop composition theories in the sixties and early seventies. The term *rhetorical* stands for inquiry that proceeds largely by deduction and analogy, that starts with probable theoretical premises, examines these premises, posits new theory derived from the premises, and argues for its viability. In the following section we will compare rhetorical research with empirical inquiry.

Rhetorical and empirical research

Several questions can be asked about these two modes.

  * What kinds of acts constitute the processes at work in them?
  * Are they complementary or conflicting?
  * What are the educational implications of this duality of design for newcomers to the field?

Some of these questions have been addressed by other composition specialists including Emig (1983), Bereiter and Scardamalia (1983), Beach and Bridwell (1984), Kaufer and Young (1983), and Winterowd (1985).

Upon close examination, we find that rhetorical and empirical modes appear to have fundamental similarities and only minor differences. Both share the same area of investigation, a domain that Phelps (1986) has defined as twofold: (1) written discourse as a complex social process by which discoursers co-construct meanings in a context and (2) the facilitation of the growth of literacy. They function, in fact, on a continuum of probable levels of knowing and interpretation. Let us examine this claim.

RHETORICAL INQUIRY

To explain the nature of rhetorical inquiry we use here as examples the work of three composition theorists—Moffett (1968), Kinneavy (1980), and Young, Becker, and Pike (1970). In what kinds of investigative processes did they engage? Each one started with a *motivating dissatisfaction*. Moffett was bothered by a disparity between the emphasis on English as a content field and studies of children's cognitive development. Kinneavy was troubled by a confusion between aims and modes of discourse. Young and colleagues were concerned about the lack of an "art" of invention. These irritations were motivating because they did not turn into free-floating anxiety, but instead were transformed into catalysts for inquiry, into *questions* that specified directions for research, that pointed out what was needed to eliminate these perceived inadequacies.

To answer these questions, these theorists went to other disciplines, looking for ways in which similar problems had been addressed. Moffett found Langer's notion of structure and Piaget's theories of cognitive development. Kinneavy turned to semiotics. Young and co-workers studied the inquiry processes of linguists, scientists, and artists. In other words, they used work in other fields as *heuristics*, as analogies to help them go beyond the known.

These acts—turning puzzlement into questions and heuristic thinking—were only preparatory. The core of their investigation entailed *creating a new theory* and *arguing for its adequacy and advantages* over existing explanations. The heart of Moffett's inquiry was his new developmental curriculum and his arguments for its power to facilitate individual growth in the language arts. The essence of Kinneavy's inquiry was a new classification of discourse and warrants for the advantages of this system. The core of Young's work was a new art of rhetorical invention based on tagmemic linguistics and a defense of its adequacy and usefulness. The creation of a new theory shares the features of any creative act—an interaction between the context, heuristics, intuition, and the constraints of the person and the field. The justification of a new theory is essentially a rhetorical act, an act of interpretation, an act of providing warrants, good reasons, of detailing the components of the theory under scrutiny, its properties, its parts and their interrelationships, and the larger groups in which it exists. Justification demands reasons why this theory is sufficient to explain the majority of instances of the behavior, and proof that the theory is not fraught with serious objections.

Rhetorical inquiry, then, entails several acts: (1) identifying a motivating concern, (2) posing questions, (3) engaging in heuristic search (which in composition studies has often occurred by probing other fields), (4) creating a new theory or hypotheses, and (5) justifying the theory.

EMPIRICAL RESEARCH

Most empirical research is initiated in the same way as rhetorical—by a motivating sense of anomaly and the posing of questions. Research-

ers notice problems in their classrooms or compelling gaps and inconsistencies in the research literature. To investigate these questions, they
use inductive processes instead of the deductive and analogical processes of rhetorical inquiry. Inductive processes take two forms: descriptive and experimental. The final act of an empirical study is very
similar to that of rhetorical research—it is essentially one of interpretation and argument for the meaning, the significance both of the problem and the results. The data and results do not speak for themselves.

Relationships between empirical and rhetorical inquiry

Empirical research, therefore, has deep similarities with rhetorical inquiry. Both types of research share the same problem domain; both
entail similar problem formulation, and both create new insights. Although each uses different search processes—rhetorical inquiry turns
to other fields analogically and empirical research amasses, analyzes,
interprets data—they both argue for the significance of their problems
and conclusions. These arguments to establish the status of new work
entail attributions of consensus that are characteristic of social knowledge. Both types of research also address two kinds of audiences: the
epistemic court of experts in the field and larger affected populations.
The notions of attributions of consensus and two types of audience are
explained further in "Composition Studies: Dappled Discipline" (Lauer,
1984b).

Rhetorical and empirical studies not only have deep similarities but
also interact. Rhetorical inquiry suggests behaviors, environments, or
populations for empirical study. It prompts coding schemes, survey
categories, and evaluative criteria. It provides hypotheses for experimental research. In return, empirical research refines rhetorical theory,
helps verify or repudiate it, and identifies important variables that contribute to new theory formation.

Advantages of multimodality

The use of these two types of inquiry, as well as historical, linguistic,
and philosophical studies, presents the field of composition studies with
a mixed blessing. Multimodality helps it avoid the nearsightedness that
causes some other fields to overlook major problems because they fall
outside the domain of a particular mode of inquiry. Such fields work
in shrinking circles. Multimodality also encourages a fruitful reciprocity

among modes of inquiry. Historical research traces and reinterprets the sources of beliefs, practices, and problems. Rhetorical theory guides empirical research, which in turn helps verify theory, a reciprocity that other fields lack. Some humanities disciplines lose touch with human affairs, while some social sciences become mesmerized by good instruments to the point that scientists seek things to measure with these tools.

But multimodality has its hazards. Keeping up with the literature becomes a challenge for those already in the field. It poses even greater problems for graduate students in composition and those in English and education departments just becoming interested in composition studies.

Such problems have no easy solutions. An unfortunate and simplistic answer would be to privilege one method of inquiry, valuing, for example, only that knowledge that can be derived empirically. Another troublesome solution would be to encourage individuals to choose among the modes of inquiry, maintaining literacy in only one, an answer that would lead to parallel but separate strands of inquiry. We contend that deliberate and interactive multimodality, especially rhetorical and empirical research, offers a richer opportunity for studying the complex domain of composition studies.

General principles of empirical research

Empirical research is the process of developing systematized knowledge gained from observations that are formulated to support insights and generalizations about the phenomena under study. Knowledge develops in empirical fields from the establishment, classification, organization, and interpretation of facts. All the research designs treated in this book are governed by four general principles used by the psychological and social sciences to develop their knowledge. The first principle is that no knowledge can be gained without comparison. This principle comes from Boring (1954) and was further explicated in Campbell and Stanley's (1963) well-known monograph. In science, the definition of a "fact" is a statement of a perceived relationship between at least two variables under study. A *variable* is a scientific name used to identify a property or dimension of the object or function under study. To learn the meaning of facts, the significance of data, researchers need a context of comparative relationships. Comprehending the nature of something involves understanding one's observations, which in turn requires comparisons.

The second principle is that measurement instruments are imprecise. They are not necessarily biased nor incorrect, just somewhat vague. A conclusion is usually stated in terms of the variance it explains. *Variance* represents the degree of individual differences among people on a variable such as writing anxiety or syntactic fluency. The precision of a measurement instrument is indicated by its reliability, which is rarely as high as .95. Even such a high level of precision only explains 90.25 percent of the variance in the behavior examined, a ratio of "error" to "truth" of about 1:9. This imprecision must always be kept in mind when a researcher presents conclusions and a reader accepts them.

A third principle is that an applied field, as Cook and Campbell (1979) indicate, is concerned with practical *decisions:* choosing alternative pedagogies, environments, and curricula. Such decisions inevitably involve making judgments about causality, judgments research can bolster.

A fourth principle is that the psychological and social sciences have a relatively large number of basic dimensions of human behavior, on the order of 400 (Nunnally, 1978) or perhaps several hundred more in contrast to the physical sciences, which have only a few basic dimensions like temperature, mass, time, space, and charge. These 400 or more dimensions vary markedly among people and cannot be held constant; people and organizations change over time. Attempting to keep under observation that many variables is a challenging task and subject to many problems.

The genesis of research

PROBLEM FORMULATION

Researchers initiate empirical research in a variety of situations: when they notice compelling problems in the classroom, gaps or inconsistencies in the literature, or incongruities between cultural expectations for writing and its present status. This sensitivity to problems and the ability to translate it into productive investigation is a trait that often separates the outstanding researcher from the average one. Many composition instructors face problems, but they often do not know how to formulate them into research designs. Instructors who see problems in practice must be able to specify the elements that are in conflict or are puzzling. In articulating these elements, researchers already begin to abstract from the flow of ongoing events. They have, in fact, initiated investigation.

For problem formulation to be effective, the questions posed must be compelling enough to sustain extended and arduous inquiry. They must also fit the investigator's research capabilities and, if answered, must make a contribution to the field, falling within the realm of those problems deemed significant and unresolved.

LITERATURE SEARCH

In order to know which problems are unresolved, a researcher must be aware of the literature in the field. A literature search assures investigators and eventually their readers that the inquiry is not redundant, that it advances an important strand of research in the field, and that it coheres within that strand. The studies cited at the end of each chapter are not organized according to the kinds of problems they address, but if readers examine carefully each study's problem statement and literature search, they will develop a knowledge of the research network already in place around many important dimensions of writing. See Conway and McKelvey (1970) for more ideas on the role of relevant literature.

Constructs: conceptual and operational definitions

Empirical research in composition seeks to develop theories that explain writing, its contexts, acquisition, and facilitation. A theory is an intellectual tool, a conceptual framework that guides empirical research and helps an investigator to understand and explain behavior. It is a set of generalizations drawn from facts, which are statements about interrelationships among variables. No theory in empirical research can exist without facts; interpretations of facts are difficult without theory. Theories are made up of *constructs*, the term for an abstract label of some element of behavior, for example planning, revising, audience adaptiveness, syntactic fluency, and so forth. The term *construct* conveys the idea of a created entity constituting many features. Cronbach and Meehl (1955) call a construct a "postulated attribute of people" (p. 283) that a researcher assumes will exist and operate in certain situations and can be distinguished from other constructs by a conceptual definition. Nunnally (1978) explains that to the extent that a variable is abstract rather than concrete, it is a construct—put together from someone's imagination. It represents the hypothesis that varieties of behaviors will correlate with one another. He goes on to say that all scientific theory concerns constructs rather than specific observable variables (pp.

96–97). A construct, therefore, is an inductive summary that attempts to characterize a facet of behavior.

A *conceptual definition* provides the essential distinguishing features of a construct. An *operational definition* specifies how a construct will be observed or measured, telling others in the discipline a minimum of things to do or look for in order to identify the construct. To create conceptual and operational definitions, a researcher can specify dynamic characteristics (what it does, how it operates, and what its active properties are), static characteristics (what it looks like), and operations that would be necessary to create or measure it. Cronbach and Meehl (1955) contend that defining a construct entails setting forth conditions in which it occurs, or setting up a network around it that explains (1) how its observable properties relate to each other, (2) how it relates to other observables, and (3) how it relates to other constructs (pp. 290–291). When formulating operational definitions, researchers are usually guided by the overall theory they hold. Moreover, a construct will probably have a number of conceptual and operational definitions given by various researchers. These multiple definitions combine so that eventually they more fully describe and define the construct.

To illustrate these important terms, we will examine several constructs and some conceptual and operational definitions developed by composition researchers to explain the behavior of invention. Both rhetorical and empirical modes of inquiry have contributed to this effort. To exemplify these definitions, we will use the work of rhetorical theorists Rohman and Wlecke (1964) and Young, Becker, and Pike (1970), and of empirical researchers Emig (1971) and Flower and Hayes (1981a, b), all of whom have been developing various constructs to explain invention.

Rohman and Wlecke identified a construct they called "prewriting" and conceptually defined it as (1) anything before the "writing ideas" are ready for the words and the page; (2) a stage of discovery in the writing process when a person assimilates his subject to himself; (3) the kind of thinking that brings forth and develops ideas, plans, and designs; (4) the imposition of pattern upon experience; (5) the finding of an arrangement that fits the subject and self; and (6) the discovery of what to say and a personal context. They also articulated some broad operational definitions, specifying that prewriting could be "imitated" and observed in three ways: keeping a journal, practicing principles derived from religious meditation, and creating analogies.

Young, Becker, and Pike identified two constructs under what they termed "preparing." The first—identifying and stating the problem— they conceptually defined as a receptivity and sensitivity to problematic situations, a defining of the problem adequately, and a stating of the unknown. Their operational definition included (1) making an explicit description of the problematic situation, (2) identifying its similarities to previous problems, and (3) classifying the unknown in different categories of questions. The second construct—exploring the problem—they conceptually defined as examining the problem from different perspectives in order to stimulate intuition and prepare for a workable solution. Their operational definition specified a conscious and systematic questioning of the problem, guided by the tagmemic heuristic procedures. Other rhetorical theorists, both contemporary and historical, have formulated alternative conceptual and operational definitions for invention (Lauer, 1984a).

Emig differentiated two constructs: prewriting and planning. She operationally defined prewriting as "the part of the composing process that extends from the time a writer begins to perceive selectively certain features of his inner and outer environments with a view to writing about them—to the time when he first puts words or phrases on paper elucidating that perception" (p. 39). She also specified the content of prewriting—a consideration of field and mode of discourse. She operationally defined planning as "any oral and written establishment of elements and parameters before or during a discursive formulation" (p. 39).

Flower and Hayes (1981a, b) defined the construct "planning" as the cognitive process in which writers form an internal representation of the knowledge that will be used in writing. They operationally defined several subprocesses: (1) generating ideas (retrieving relevant information from long-term memory), (2) organizing (giving meaningful structure to ideas, identifying categories, searching for subordinate and superordinate ideas, and attending to textual decisions about presentation and ordering), (3) goal setting (creating procedural and substantive goals), and (4) pausing at the beginning of composing episodes (global rhetorical planning and problem solving instead of generating what to say next). Table 1-1 illustrates these definitions.

Note that conceptual definitions are more abstract than operational definitions, which indicate ways in which the variable operates and can be observed. No clear line separates these two kinds of definitions— they exist on a continuum. Notice also that the sources of these definitions differ. Rohman and Wlecke and Young, Becker, and Pike gleaned their variables from theories of creativity, inquiry, problem solving, cognitive dissonance, tagmemics, and so forth, and from their experiences as writers and teachers. Emig and Flower and Hayes developed their definitions from direct observations of writers, from theories of

Table 1-1 Constructs: Conceptual and Operational Definitions

Researchers	Construct	Conceptual definition	Operational definition
Rohman and Wlecke (1964)	Prewriting	1. Anything that goes before writing ideas are ready for the page 2. Stage of discovery when person assimilates subject to self 3. Kind of thinking that brings forth and develops ideas, plans, designs 4. Imposition of pattern on experience 5. Finding arrangement that fits subject and self 6. Discovering what to say and personal context	Prewriting can be imitated and observed by 1. Keeping a journal 2. Practicing principles from religious meditation 3. Creating analogies
Young, Becker, and Pike (1970)	Preparing; identifying and stating problem	1. A receptivity and sensitivity to problematic situations 2. Defining a problem adequately 3. Stating an unknown	1. Making an explicit description of the problematic situation 2. Identifying similarities to previous problems 3. Classifying the unknown in different categories of questions
	Exploring the problem	Examining the problem from different perspectives to stimulate intuition and prepare for a workable solution	A conscious and systematic questioning of the problem, guided by heuristic procedures

Emig (1971)	Prewriting	Content: a consideration of field and mode of discourse
		Part of the process extending from time writers perceive selectively certain features of inner and outer environments with a view to writing about them—to time they put words on paper
	Planning	Any oral and written establishment of elements and parameters before or during discursive formulation
Flower and Hayes (1981 a,b)	Planning	Cognitive process in which writers form an internal representation of the knowledge to be used in writing
		1. Generating ideas (retrieving relevant information from long-term memory)
		2. Organizing (giving meaningful structure to ideas, identifying categories, searching for superordinate and subordinate ideas, attending to textual decisions about presentation and ordering)
		3. Goal setting (creating procedural and substantive goals)
		4. Pausing at the beginning of composing episodes (global rhetorical planning and problem solving)

creativity and problem solving, and from their own experiences as writers and teachers.

CONSTRUCT VALIDATION

The validity of a construct—in other words, the extent to which it accurately and adequately accounts for the phenomenon it represents—can be determined in several ways, as explained by Cronbach and Meehl. First, researchers must operationally define a variable so that it can be observed and measured: counted, timed, videotaped, verbally described, tested, and so forth. They then can examine this variable in several ways. They can study whether groups expected to differ on the construct do indeed differ or whether groups that should not differ do not differ. For example, they can test whether experienced and novice writers differ in the amount and kind of planning they do. Researchers can also correlate the results of the measures they use with those of other measures of the construct or correlate the behaviors subsumed under the construct. For example, Flower and Hayes could determine whether the subbehaviors of organizing—giving meaningful structure to ideas, identifying categories, searching for subordinate and superordinate ideas, and attending to textual decisions about presentation and ordering—correlate with each other. Investigators can also determine whether the construct remains stable over different occasions and times as Witte and Davis did for the T-unit (1980, 1982) or to what extent intervention affects the construct. Young and Koen (1973) and Odell (1974) conducted this kind of construct validation by introducing treatments to see if the constructs of identifying, stating, and exploring problematic situations could be enhanced. They operationally defined the constructs and measured changes.

Campbell and Fiske (1959) describe construct validity as a response tendency that can be observed under more than one condition, can be meaningfully differentiated from other traits, and can sustain more than one operational definition. In other words, as Cronbach (1971) explains, construct validation requires the integration of many studies that examine the same or highly related constructs and test every link in the network surrounding the construct. Hypotheses and counter-hypotheses about the definitions of the constructs must be advanced and tested. Researchers must also check the influences on the constructs, the intercorrelations among features within a construct, and the distinctions between the construct and others.

As conceptual and operational definitions increase and converge, the construct becomes richer and better understood. In fact, the process of construct validation is an integral part of the process of empirical research, which has the following stages: the development of constructs, their operational definition, their measurement and analysis, and their generalization. Each of the subsequent chapters explains how this effort entails more rigorous quantification, measurement, and statistical analysis.

Research taxonomy

We organize the text using the taxonomy of research designs presented in Figure 1-1, a taxonomy that stems from the axiom and principles above. Our order of presentation moves cumulatively from less to more quantitative and statistical explanation. For that reason, we introduce the principles of measurement including reliability and validity in Chapter 7, at a midpoint between descriptive and experimental research. These principles, however, govern all research designs. The glossary and appendix will help readers who need further explanations of measurement. We discuss program evaluation in the last chapter because it makes use of all the previous designs.

DESCRIPTIVE RESEARCH

Descriptive studies entail observation of phenomena and analysis of data with as little restructuring of the situation or environment under scrutiny as possible. Several kinds have been used in composition research: *case studies*, focusing on one or a few individuals; *ethnographies*, examining environments; *surveys*, seeking information about larger groups usually by means of sampling techniques; *quantitative descriptive studies*, quantifying and analyzing interrelationships among variables; and *prediction and classification studies*, gathering and analyzing the records of individuals in order to predict future behavior or similarities with certain groups.

Although some of these types of descriptive research entail complex statistical analyses, researchers do not deliberately structure or control the environment from which the data are gathered. Descriptive research seeks to further develop theory because researchers are trying to isolate and explain important variables. After researchers visit or interview subjects, take notes, and make protocols, they face the task of coding, of sorting the information into categories, of noticing patterns and relationships—in short, of interpreting. But where does a coding system come from? It does not spring directly from the data. Rather it

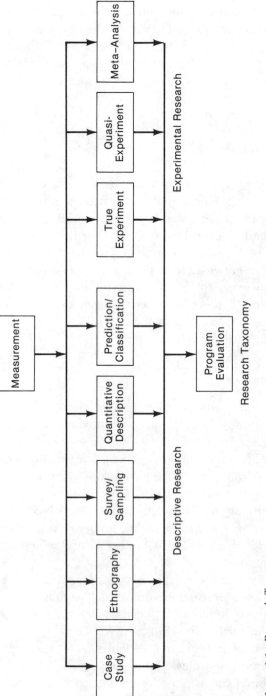

Figure 1-1 Research Taxonomy.

arises from an interplay between the data and the researcher's knowledge of theory. For example, Flower and Hayes's coding (1981a) comes from a reciprocity between the theory of problem solving, their views of writing, and the protocols of their subjects. In survey research, theory, whether tacit or explicit, guides the development of items in a questionnaire. In prediction and classification studies, theories of what constitutes regular or remedial instruction influence the results. Thus, theory can be said to drive descriptive studies.

The status of this kind of empirical research varies from field to field. Some behavioral and social scientists view it as prescientific, to be done at earlier stages of an investigation when one is looking for hypotheses to test. Fields like ethnology and anthropology, however, give it a high status. Right now in composition studies, a number of researchers like Emig argue for a high priority for descriptive research.

EXPERIMENTAL RESEARCH

Another kind of empirical research, experimental, is highly valued in many social sciences because the data from such research yield better evidence of cause-and-effect relationships. Such research is also theory-driven in that it begins with hypotheses whose sources are usually theoretical. To test these, researchers rearrange the environment into treatment and control groups, administer treatments, and assess the results with measurement instruments and observations that they strive to make reliable and valid.

Two types of experimental research, the true experiment and the quasi-experiment, vary in the kind of control imposed. In the *true experiment*, subjects are allocated to groups by randomization, which permits the researcher to consider the groups initially "not unequal," a major advantage because if the treatment group performs significantly better, the treatment is said to cause the results. In a *quasi-experiment*, because the groups are initially intact (i.e., not randomized), pretests and more sophisticated statistical methods must be used to establish the initial relative equality of the groups in order later to determine possible cause-and-effect relationships. In applied research, experimental studies often function to test the effectiveness of pedagogical methods.

To extend and generalize conclusions about causality, a new type of study, *meta-analysis*, examines the cumulative effectiveness of many experimental studies. *Program evaluation* uses both descriptive and experimental designs to establish and apply criteria to writing programs in either formative or summative ways. Here too the researcher's theoretical beliefs about effective composition instruction and curricula affect the choice of criteria.

But the strength of the studies, as is true of all designs, rests on the reliability and validity of the measurement instruments used. These two essential features of all research are discussed in detail in Chapter 7 and Appendix I; each chapter also treats them in relation to the design being examined.

INTERPRETATION AND RESEARCH DESIGN

We have categorized these research designs based on the principles of empirical research discussed above: comparison, precision of measurement, causality, and control of variables. In the experimental category, *true experiments* yield the most rigorous comparisons, giving a high level of probability of cause-and-effect relationships, because they bring under control all 400 or so variables intrinsic to conducting research with people in natural settings. Next in cause-and-effect power are *strong quasi-experiments*, which possess many of the features of the true experiment but do not provide as much control of the human and environmental variables. They also require more complex statistical procedures. *Weak quasi-experiments*, research studies with initially unequal groups, are more difficult to interpret because they are even more vulnerable to the inherent problems of the behavioral and social sciences—imprecise measurement and instability of human behavior.

In the descriptive category, *prediction* and *classification studies* generally do not (or should not) make cause-and-effect statements but should be used to select the best indicators of a person's future performance or of relationships among variables. When conducting these studies, researchers must understand the limits of the design, because misuse of the regression equations and even differences in the length of measures can greatly distort the results and interpretations, becoming detrimental to results. *Sampling studies* describe large populations efficiently, but it is difficult to make cause-and-effect statements from their results. *Quantitative descriptive* research studies establish statistically significant relationships among variables and the amount of variance explained by different variables. But their results are hard to use to establish cause-and-effect relationships because many additional variables are not explicitly controlled.

Conducting qualitative descriptive research, whether *case study* or *ethnography*, is a prerequisite to all types of experimental research. Unless the important dimensions and concepts of a field have been identified and even measured, their explicit cause-and-effect relationships cannot be tested in true or quasi-experiments. Much *informal* qualitative research goes on in composition classrooms as teachers identify important powers, skills, conditions, and pedagogies that need attention; formal descriptive studies must move such observations into coding and quantifying.

Bourdieu (1977), an anthropologist, argues against privileging either qualitative or experimental research. He refers to qualitative research as *subjectivist:* the ethnomethodologist has "presuppositions inherent in the position of an outside observer, who in his preoccupation with *interpreting* practices, is inclined to introduce into the object the principles of his relation to the object, as is attested by the special importance he assigns to communicative functions" (p. 2). This kind of researcher "inflicts on practice a much more fundamental and pernicious alteration which, being a constituent condition of the cognitive operation, is bound to pass unnoticed: in taking up a point of view on the action, withdrawing from it in order to observe it from above and from a distance, he constitutes practical activity as an *object of observation and analysis, a representation*" (p. 2). Bourdieu does not criticize ethnomethodological research *per se,* but only the claim that its descriptions of the environment are "equal to" the environment itself.

Bourdieu also criticizes this same tendency in experimental research to consider its models equal to the behavior studied. He terms experimental research *objectivist* and *idealist.* Here again he does not devalue this kind of research but argues that objectivist claims to identify structures, relationships, and causalities can be made only by breaking with primary knowledge and grasping "practice from outside, as a *fait accompli*" and by constituting the world as a "system of objective relations independent of individual consciousness and wills" (p. 4).

He argues for a theory of practice as a way of escaping from the ritual choice between objectivism and subjectivism. A theory of practice inquires into "the mode of production and functioning of the practical mastery which makes possible both an objectively intelligible practice and also an objectively enchanted experience of that practice" (p. 4). (For a full explanation of a theory of practice, see Bourdieu's *Outline of a Theory of Practice.*) A theory of practice entails self-reflection on the part of researchers into their own practices of research, whether qualitative or experimental, and a recognition that these kinds of research do not constitute a hierarchy of knowledge but rather are partial representations or perspectives that are adequations but not equal to primary experience.

Our text begins with qualitative research designs and moves toward experimental research, not to suggest a privileged order of designs but to show increasing use of quantification in the designs.

Illustrative studies

To clarify the research design in each chapter, we introduce two or three composition studies. After a brief summary of each study, we use it to explain the principles, concepts, and procedures inherent in the design under examination. We intend neither to analyze these studies as an end nor to propose them as representative of the range of research areas in composition. Their purpose is to illustrate features of the designs.

Even though we point out weaknesses in some of these studies, we chose them because they provide helpful examples in a field that has only recently undertaken extensive empirical research. Each study has made a contribution to the field's developing understanding of writing, examining an important composition problem. Even though some of the designs now seem inappropriate or inadequately conducted, the investigators' efforts helped other researchers to advance.

Format

We organize our discussion of each research design by explaining the following procedures: selection of subjects, formulation of hypotheses or questions, data collection, data analysis, variable identification, and conclusions. These procedures characterize both the processes of conducting empirical research and the important features to examine when reading research studies. We judge it neither possible nor desirable to avoid using the technical terms that characterize empirical research, but we discuss statistical analyses only to the extent necessary to clarify the designs. Appendix I offers a fuller treatment of the major statistical analyses used in composition research.

At the end of every chapter, we list questions to guide the reading of studies employing the design under discussion. We also list references to provide further explanation and to identify composition studies that use the design. A glossary is appended to assist readers with terminology that has been cumulatively introduced throughout the book.

References

Asher, J.W. (1976). *Educational research and evaluation methods*. Boston: Little, Brown.

Beach, R., & Bridwell, L. (Ed.). (1984). *New directions in composition research*. New York: Guilford Press.

Bereiter, C., & Scardamalia, M. (1983). Levels of inquiry in writing research. In P. Mosenthal, L. Tamor, & S. Walmsley (Eds.). *Research on writing: Principles and methods*. New York: Longman.

Boring, E.G. (1954). *The nature and history of evaluation methods*. Boston: Little, Brown.

Braddock, R., Lloyd-Jones, R., & Schoer, L. (1963). *Research in written composition*. Urbana, Ill.: National Council of Teachers of English.

Bourdieu, P. (1977). *Outline of a theory of practice*. Tr. R. Nice. London: Cambridge University Press.

Campbell, D.T., & Stanley, J.C. (1963). Experimental and quasi-experimental designs for research. In N.L. Gage (Ed.). *Handbook of research on teaching*. Chicago: Rand McNally.

Campbell, D.T., & Fiske, D.W. (1959). Convergent and discriminant validation by the multitrait–multimethod matrix. *Psychological Bulletin, 56,* 81–105.

Conway, J.A. & McKelvey, T.V. (1970). The role of the relevant literature: A continuous process. *Journal of Educational Research, 63,* 407–413.

Cronbach, L.J. (1971). Test validation. In Robert Thorndike (Ed.). *Educational Measurement*. Washington, D.C.: American Council on Education, pp. 443–507.

Cronbach, L.J., & Meehl, P.E. (1955). Construct validity in psychological tests. *Psychological Bulletin, 52,* 281–302.

Cook, T.D., & Campbell, D.T. (1979). *Quasi-experimentation: Design and analysis issues for field settings*. Boston: Houghton Mifflin.

Dijk, T.A. van (Ed.). (1985). *Handbook of discourse analysis*, Vols. 1–4. London: Academic Press.

Emig, J. (1971). *The composing processes of twelfth graders*. Urbana, Ill.: National Council of Teachers of English.

Emig, J. (1983). Inquiry paradigms and writing. In D. Goswami & M. Butler (Eds.). *The web of meaning*. Upper Montaclair, N.J.: Boynton/Cook Publishers.

Faigley, L., et al. (1985). *Assessing writers' knowledge and processes of composing*. Norwood, N.J.: Ablex.

Flower, L., & Hayes, J.R. (1981a). A cognitive process theory of writing. *CCC, 32,* 365–387.

Flower, L., & Hayes, J.R. (1981b). The pregnant pause: An inquiry into the nature of planning. *RTE, 15,* 229–243.

Hillocks, G. (1985). What works in teaching composition: A meta-analysis of experimental treatment studies. *American Journal of Education, 93,* 133–170.

Hillocks, G. (1986). *Research on written communication: New directions for teaching*. Urbana, Ill.: National Council of Research in English.

Kaufer, D., & and Young, R.E. (1983). Literacy, art, and politics in departments of English. In W. Horner (Ed.). *Composition and literature*. Chicago: Univ. of Chicago Press.

Kinneavy, J. (1980). *A theory of discourse*. New York: Norton.

Lauer, J.M. (1984a). Issues in rhetorical invention. In R.J. Connors, L.S. Ede, & A.A. Lunsford (Eds.). *Essays on classical rhetoric and modern discourse*. Carbondale, Ill.: Southern Illinois Univ. Press.

Lauer, J. (1984b). Composition studies: Dappled discipline. *Rhetoric Review, 3,* 20–28.

Moffett, J. (1968). *Teaching the universe of discourse*. Boston: Houghton Mifflin.

Moran, M.G., and Lunsford, R.E. (1984). *Research in composition and rhetoric*. Westport, Conn.: Greenwood Press.

Nunnally, J.C. (1978). *Psychometric theory*, 2nd edition. New York: McGraw-Hill.

Odell, L. (1974). Measuring the effect of instruction in prewriting. *RTE, 8,* 228–240.

Phelps, L. (1986). The domain of composition. *Rhetoric Review, 4,* 182–195.

Phelps, L. (1983). Possibilities for a post-critical rhetoric: A parasitical preface. *Pretext, 3–4,* 201–213.

Rohman, D.G., & Wlecke, O. (1964). *Prewriting: The construction and application of models for concept formation in writing*. Cooperative Research Project #2174, Cooperative Research Project of the Office of Education, U.S. Department of Health, Education, and Welfare.

Scardamalia, M., & Bereiter, C. (1986). Research on written composition. In M.C. Wittrock (Ed.). *Handbook of research on teaching*, 3rd edition. New York: Macmillan.

Smith, J.K. (1983). Quantitative versus qualitative research: An attempt to clarify the issue. *Educational Researcher, 12,* pp. 6–13.

Vitanza, V. (1987). Critical sub/versions of the history of philosophical rhetoric. *Rhetoric Review, 6,* 41–66.

Winterowd, R.W. (1985). The politics of meaning. *Written Communication, 2*(3), 269–291.

Witte, S. & Davis, A. (1980). The stability of T-unit length: A preliminary investigation. *RTE, 14,* 5–17.

Witte, S. & Davis, A. (1982). The stability of T-unit length in the written discourse of college freshmen: A second study. *RTE, 16,* 71–84.

Young, R.E., Becker, A., & Pike, K. (1970). *Rhetoric: Discovery and change*. New York: Harcourt Brace Jovanovich.

Young, R.E., & Koen, F. (1973). *The tagmemic discovery procedure: An evaluation of its uses in the teaching of rhetoric*. University of Michigan, Ann Arbor. National Endowment for the Humanities Grant No. EO 528-71-116.

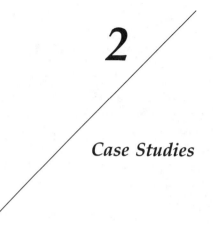

2

Case Studies

Qualitative descriptive research is often the first kind of empirical research done in fields that seek to identify the important aspects or variables of any phenomenon to be studied. In composition classrooms, instructors engage in informal observation all the time, noticing their students' behaviors, the results of instruction, the difficulties that occur, the methods that seem to work, and so forth. Every composition classroom exhibits hundreds of aspects of composition. But which of these are important? Which of these deserve close scrutiny and rigorous analysis? Which are key variables of composition?

Qualitative descriptive research tries to answer these questions by closely studying individuals, small groups, or whole environments. It tries to discover variables that seem important for understanding the nature of writing, its contexts, its development, and its successful pedagogy. When researchers engage in descriptive research, they examine and analyze segments or whole situations as they occur. This kind of research, therefore, does not primarily attempt to establish cause-and-effect relationships among variables; it seldom has that kind of explicit power. It is, instead, a design that, by close observation of natural conditions, helps the researcher to identify new variables and questions for further research. Researchers in composition generally find variables to study from their own classroom experience, from rhetorical inquiry, from previous empirical research on composition, or from variables identified in other fields such as psychology, sociology, or linguistics. The type of descriptive research that this chapter will discuss is case study. We will use three well-known studies to explain this design.

Three case studies

Study 1   Janet Emig (1971). *The composing processes of twelfth graders.* Urbana, Ill.: NCTE.

In 1971, Janet Emig published the results of a case study she conducted with eight twelfth graders (five girls and three boys) in the Chicago area. Emig wanted to study the composing processes of these students as they produced two different kinds of writing that she called reflexive (self-sponsored) and extensive (school-sponsored). She conducted four interviews with each student to collect various data from them. She then analyzed these data and published the results. We will discuss this study in more detail as the chapter progresses.

Study 2   Donald Graves (1975). An examination of the writing processes of seven-year-old children. *RTE, 9,* 227–241.

From December 1972 to April 1973, Donald Graves studied the writing processes of eight seven-year-old subjects (six boys and two girls) from four classrooms (two informal and two formal). After observing them both inside and outside school, interviewing and testing them, and examining other related variables, Graves analyzed the data he had collected, identified several salient variables, and published his conclusions. We will discuss this study in more detail below.

Study 3   Linda Flower and John Hayes (1981).
The pregnant pause: An inquiry into the nature of planning.
RTE, 15, 229–243.

This study is one of a number of descriptive studies of the writing process conducted by Flower and Hayes. They examined the thinking processes of four subjects (three expert writers and one novice), in order to determine the content of pauses between "composing episodes." The design is discussed in the following paragraphs.

Research design: case study

Subject selection

Those people a researcher selects to examine are called *subjects*. Emig chose eight volunteers who attended a variety of types of schools: an all-white upper-middle-class suburban school, an all-black ghetto school, a racially mixed lower-middle-class school, an economically and racially mixed good school, and a university school. Intelligence tests for five of the students revealed that three were above average and two were average. According to their teachers and National Council of Teachers of English (NCTE) Achievement Awards, six were judged good writers and two average. Based on teacher recommendations, Graves selected eight students, two from each of four classrooms that were being observed in a larger study. These children were judged "normal" by their instructors. Flower and Hayes introduced another element into their subject selection, the distinction between expert and novice that was common in the problem-solving research underlying their methodology. They chose three experts and one novice writer. In the report of their study, however, they gave no other information about their subjects or about the nature of the writing task. Such information can be found in reports on some of their other research that uses the same subjects and task.

Hypotheses and background theory

Researchers bring some theory to their research. Theory is sometimes expressed in the form of hypotheses or questions articulated at the start of the research or at the beginning of the published report. Even with hypotheses or questions in mind, however, a descriptive researcher tries to withhold judgment in order to allow the weight of data to suggest new conclusions. Emig began with a theory about writing as a process. She had several hypotheses: that writing in the two modes (reflexive and extensive) would result in different lengths of discourse and different clusterings of the components of the writing process, that she could glean these variations by listening to students composing aloud, and that these differences would manifest themselves in an explicit set of stylistic principles. Her opening questions later helped her code her data.

Graves was also interested in the writing process, conjecturing that the types of learning environment (formal or informal) might influence the composing processes of children, a speculation derived from his experience and larger research efforts. Flower and Hayes began their study with two hypotheses to explain the content of pauses during planning and writing: the linguistic hypothesis, which held that writers pause to plan or generate what they will say next, and the rhetorical hypothesis, which held that writers pause to carry out global rhetorical planning or problem solving that is not necessarily connected to an immediate piece of text. The second hypothesis stemmed from their underlying theory of problem solving. It is important for researchers to be self-consciously critical of the theoretical assumptions underlying their research.

Data collection

In a case study, researchers collect data in a variety of ways and for different purposes. To obtain as complete a picture of her students' processes as possible, Emig gathered data from several sources: conversations, tape recordings of composing aloud and accounts of processes, discrete observations of composing, writing samples from each student, and school records. Her data included two pieces of extensive writing from each student, autobiographies from each, six accounts of the process, a piece of obscenity, and a nonresponse. Her use of several methods of data collection not only gave her a better data base for her analysis but also introduced composition researchers to new means of observing and gathering material.

Graves gathered his data by interviewing students and parents, using tests of reading and intelligence, collecting folders of assigned and unassigned writing, studying records of educational development, and observing students in school and outside of school. The use of folders, observations of 53 writing episodes (see Table 2-1) and interviews gave him extensive data about these students' writing behavior.

Flower and Hayes used *protocol analysis* to gather their data, defining a protocol as "a description of the activities, ordered in time, in which a subject engages while performing a task" (1981a). As subjects thought aloud while composing, a tape recording of this effort was made and transcribed (see Figure 2-1). The use of this tool gave Flower and Hayes a way of analyzing a part of writing behavior that had been difficult to study before. For a discussion of the validity of this instrument and directions for its use, see Perkins (1981) and Ericsson and Simon (1984).

Variable development: coding and content analysis

The most crucial task of a case study is the identification of important variables in the data. Sometimes this task is called *coding*—the setting up and labeling of categories, which then become the *variables* of the study (see Table 2-2). The broader term for this effort is *content analysis*,

Table 2-1   Examples of Interventions Made By Observer During Writing Episodes

Phase in Episode	Setting at Time of Intervention	Observer's Objective	Observer's Questions or Statements
Prewriting Phase	1. The child was about to start drawing his picture.	1. To determine how much the drawing contributes to the writing.	1. "Tell me what you are going to write about when you finish your drawing."
	2. The child has finished his drawing.	2. To determine how much the drawing contributes to the writing.	2. "Tell me what you are going to write about now that you have finished your drawing."
	3. The child has finished his drawing.	3. To determine in less direct fashion how the drawing contributes to the writing.	3. "Tell me about your drawing."
Composing Phase	1. The child is about to start writing.	1. To determine the range of writing ideas possessed before child writes.	1. "Tell me what you are going to write about."
	2. The child attempts to spell a word.	2. To determine the child's understanding of the resources available for spelling help.	"That seems to be a hard one. How can you figure out how to spell it? Tell me all the different ways you can figure out how to spell it."

Source: Graves, (1975), Table II. Copyright © 1975 by the National Council of Teachers of English. Reprinted by permission of the publisher.

which is a major measurement procedure, allowing researchers to claim that materials and observations are ultimately quantifiable. The method is designed for use with communication data of all kinds: essays, television shows, presidential speeches, letters, and so forth. Researchers analyze the communication data, notice patterns, identify and operationally define variables, and relate them to one another. Once researchers have identified the categories or variables, they must test them for reliability (see Appendix I) by developing instructions that will enable independent judges to conduct content analyses on representative data.

Krippendroff (1980, p. 174) advises that coding instructions should contain

- A prescription of the characteristics of the observers employed in the recording process
- An account of the training of these observers

Episode 1   My Job[2] for a young—[2]Oh I'm to describe my job for a young
thirteen to fourteen year-old teenage female audience—Magazine
[3]
—Seventeen. -a- My immediate reaction is that it's utterly impossible. I did read Seventeen, though—I guess I wouldn't say I
[4.2]
read it -a- I looked at it, especially the ads, so the idea would be
to describe what I do to someone like myself when I read—
[2]
well not like myself, but adjusted for—well twenty years later.
[5.2]
-a- Now what I think of doing really is that—until the coffee
comes I feel I can't begin, so I will shut the door and feel that
I have a little bit more privacy,//
[6.2]

Episode 2   -um- Also the mention of a free-lance writer is something I've
—I've no experience in doing and my sense is that it's a—a
formula which I'm not sure I know, so I suppose what I have to
[2]
do is -a- invent what the formula might be, and—and then try
[3.8]   [4.2]
to -a- try to include—events or occurrences or attitude or ex-
[3.6]
periences in my own job that would -a- that could be—that

Episode 3a   [2]
could be conveyed in formula so let's see -// I suppose one
would want to start—by writing something—that would -a-
[2]
attract the attention of the reader—of that reader and -a- I
suppose the most interesting thing about my job would be that
it is highly unlikely that it would seem at all interesting to
[2]
someone of that age—So I might start by saying something like
—Can you imagine yourself spending a day—Many days like
this—waking up at 4:30 a.m., making a pot of coffee . . .
looking around . . . my—looking around your house, letting in
your cats . . . -a- walking out—out with coffee and a book and
watching the dawn materialize . . . I actually do this . . . although
4:30's a bit early, perhaps I should say 5:30 so it won't seem—
although I do get up at 4:30 -a- watching the dawn materialize
and starting to work—to work by reading—reading the manuscript—of a Victorian writer . . . with a manuscript of a . . . a
Victorian writer . . . a person with a manuscript of a student
—Much like yourself—Much like—Much like -a- a student or
a book by Aristotle they've heard of Aristotle or—who could
I have it be by—Plato probably When it gets to be—When
you've . . . -a- finished your coffee and whatever you had to
do (Oh thanks)—whatever—now I've just gotten coffee—finished
your coffee (mumbling) . . . when you've finished your coffee
[3]   [2]
and -a- foreseen—and -a- ummmmmm—when you've finished your
coffee, you dress and drive—about three miles to the university

where you spend another—where you spend—you spend hours
—you spend about—oh what—four or five—supposed to be four
hours—about three hours a day—about three hours teaching—
many more hours talking to students—talking to—talking to

6.8

Episode 3b   other teachers . . . Um -/ should I (mumble)—the thing is
about saying teachers—the—the teenage girl is going to think
teachers like who she has, and professor I always feel is sort of
pretentious and a word usually—usually I say teacher, but I

2   7.6

know that means I . . . It's unfortunate now in society we
Episode 3c   don't—but that that isn't prestige occupation./ Talking to other
people like yourselves—that's whoever it may be—other people
at your job—other—other people like yourself—uh a lot like
yourself but—talking to other people like yourself—going to
meetings . . . committee meetings . . . and doing all this for nine
months so that the other three . . . and doing all this for three
months—okay—nine months . . . If you can imagine that . . .

Figure 2-1 Episodes in an expert writer's protocol. Planning—no underlining; text production—underlining; reading—double underlining; major episode boundary—double bar;
minor episode boundary—single bar; time of pauses more than two seconds—super-
script numbers. *Source:* Flower and Hayes (1981a), pp. 235–236. Copyright © 1981 by the
National Council of Teachers of English. Reprinted by permission of the publisher.

- A definition of the recording units including procedures for their
  identification
- A delineation of the syntax and semantics of the categories, including an outline of the cognitive procedures to be employed in
  placing data into categories
- A description of how data sheets are to be used and administered.

If this analysis is to be replicable, instructions must be explicit and detailed. If these instructions do not become part of the published report,
they must at least be available upon request.

What is the source of the categories developed during the analysis?
They are not in the data; researchers impose them on the data or develop them from the data, identifying dimensions, aspects, or categories that seem important for understanding writing, based on the theory and hypotheses that researchers bring to the experiment and on
their intuition and perception, interplaying with relationships and patterns seen in the data. Sometimes variables come from previous composition studies or from other fields. It is often the inventiveness and
quality of variable development and definition that distinguish a noteworthy from a poor piece of research. Important here in category development is obtaining agreement among coders using the operational
definitions.

Content analysis was tedious and time-consuming until computers

Table 2-2   Example of a Writing Episode

A whale is eating the
  1   2   3   4   5
	10:12
	R
		9—Gets up to get dictionary. Has the page
		    with pictures of animals.	

men, A dinosaur is 6 7 8 ⑨ ⑩ ⑪ 12			

triing to eat the whale.
  13   14 15 16   17   ⑱
	IU
	R
		10—Teacher announcement.
		11—Copies from dictionary and returns book
		    to side of room.	

A dinosaur is frowning ⑲ ⑳ ㉑ 22 23 ㉔			
a tree at the lion. and ㉕ 26 27 28 29 30 31 32	RR	18—Stops, rubs eyes. 19—Rereads from 13 to 19.	
the cavman too. the men 33 34 35 ㊱ 37 38	OV OV	20—Voices as he writes. 21—Still voicing.	
are killed. The dinosaur 39 40 41 42 43	IS	24—Gets up to sharpen pencil and returns.	
killed the whale. The 44 45 46 47 49 ㊽	RR RR	25—Rereads from 20 to 25. 36—Rereads to 36. Lost starting point.	
cavmen live is the roks. 50 51 52 53 54 55 ㊶		48—Puts away paper, takes out again.	
	RR 10:20	56—Rereads outloud from 49 to 56.	

Source: Graves (1975), Table 1. Copyright © 1975 by the National Council of Teachers of English. Reprinted by permission of the publisher.

Key: 1-2-3-4—Numerals indicate writing sequence. ④—Item explained in comment column on the right. ////—erasure or proofreading. T—Teacher involvement; IS—Interruption Solicited; IU—Interruption Unsolicited; RR—Reread; PR—Proofread; DR—Works on drawing; R—Resource use. Accompanying Language: OV—Overt; WH—Whispering; F—Forms letters and words; M—Murmuring; S—No overt language visible.

began to facilitate the analysis. In 1966, Phillip Stone and others developed the General Inquirer, which revolutionized content analysis. The General Inquirer is a computerized method of content and statistical analysis that has general and special "dictionaries" that search data in order to find instances of variables. Words such as *I, my,* and *mine* have been used, for example, to search for the variable of self-reference. Variables such as social status, need to achieve, and sexism have been traced. The General Inquirer has been used to content-analyze small-group interactions, presidential candidates' acceptance speeches, folktales, songs of primitive people, imaginative processes, and specificity in essays. Researchers can add to it their own dictionaries to search for the variables they wish to find in the data. Today, a number of computer programs like Writer's Workbench (Frase, 1980) have been developed to analyze texts for various features like punctuation and style (see Bridwell, Nancarrow, and Ross, 1984; Milic, 1981; Schwartz and

Bridwell 1984). Content analysis, therefore, is an exceptionally useful method of developing and quantifying variables in communication data.

AGREEMENT AMONG OBSERVERS

It is important for more than one observer to code the data in order to establish reliability. Three common ways exist to determine the level of agreement: percentage of agreement, Cohen's kappa (1960), and Cronbach's alpha. See Appendix I for descriptions of these statistics.

VARIABLE IDENTIFICATION IN THE THREE STUDIES

Emig identified a number of variables, including (1) the context of composing (community, family, school), (2) the nature of the stimulus (registers: field of discourse, mode of discourse, tenor of discourse; self-encountered; other-encountered: assignment by teacher, reception of assignment by student), (3) prewriting, (4) planning, (5) starting, (6) composing aloud, (7) reformulating, (8) stopping, (9) contemplation of the product, and (10) teacher influence (pp. 33–44). One of her contributions to composition scholarship was her identification of *process* variables and her examination of their operation in different types of discourse.

Graves selected the following variables for the eight students: (1) learning environments (formal or informal), (2) sex, (3) developmental level, (4) use of language, and (5) problem-solving behavior. For Michael, a subject who was studied intensively, Graves isolated (1) family, (2) home, (3) teacher, (4) developmental level, and (5) peer influence. His published report gave the results of his coding and a table of one episode and interview, but not his complete method of analysis. Readers can turn to his dissertation (1973) for more information.

Flower and Hayes studied the relationship among the following variables: (1) episodes (number, clauses per episode, range of clauses per episode, time), (2) episode boundaries, (3) paragraph boundaries, (4) topics, and (5) rhetorical goal setting. To help them identify episode boundaries and other variables, they used four trained judges, four intuitive judges, and twenty-two writing researchers. They calculated the percentage of agreement among their judges to be between 66 and 75%. Among the strengths of the study were the use of a large number of people to code the data, the specific tables that show readers the range of codings, and their sample of one expert's protocol.

Results

Reports of case studies usually entail extensive descriptive accounts. Conclusions can be made only about those subjects studied. Emig described her results in detail. She found that her twelfth graders engaged in both reflexive and extensive writing, which were characterized by different lengths and clusterings of components. Extensive writing was school-sponsored. Reflexive writing was a lengthier process with more components: a longer prewriting period, with more discernible moments of stopping, starting, and contemplating the product, and more frequent reformulations. Reflexive writing occurred often as poetry, with the self or trusted peer as audience. Extensive writing occurred chiefly as prose, with a detached and reportorial attitude toward the field of discourse, and with adult others and teachers as audience. For her other conclusions about the composing process, see the full report (1971), which includes a profile of Lynn, one of the subjects. Because Emig was able to publish her results in monograph form, she could describe the methods and results of her study in detail. She was careful to present her findings as conclusions that applied only to her subjects, not to writers in general. She did, however, suggest further questions, hypotheses, and future implications for research, appropriate and valuable features of qualitative studies. Her work initiated the use of the case study design to investigate the writing process at a time when neither was characteristic of research on writing (Braddock, Lloyd-Jones, and Schoer, 1963).

Graves presented his results in the form of directions and questions. About three of his variables, he reported conclusions:

1. Learning environments influenced writing: informal environments provided greater choice, required no motivation or supervision, and favored boys; formal writing favored girls. Assigned writing was shorter and inhibited the range, content, and amount of writing done.
2. Sex differences were apparent in writing: girls wrote longer products, less unassigned writing, and more first-person discourse. They also wrote less secondary and extended writing, more about home and school, and were less concerned about neatness and spacing. They stressed more prethinking and organizational qualities, explored feelings in their characterizations, and gave more illustrations to support their judgments.
3. Developmental factors produced two different kinds of writers: reactive and reflexive.
4. For Michael, many variables contributed in unique ways at any given point in his process of writing.

Graves also identified several variables that emerged as important for further study: (1) use of first and third person reported in thematic choices; (2) the secondary and extended territoriality reported in thematic choices; (3) prewriting, composing, and postwriting phases in the writing episode; and (4) components of profiles to assess children's developmental levels. He concluded with several questions (pp. 235–241) that he has used as a basis for subsequent research.

Flower and Hayes also drew several conclusions about the predictions they set at the beginning.

> *Prediction 1:* Paragraphs were poor predictors of major episodic boundaries.
>
> *Prediction 2:* Logical topics, broadly defined, were very effective in categorizing most of the material writers produced, but they were no better in detecting episode boundaries than paragraphs had been.
>
> *Prediction 3:* Planning and pausing that occurred at episode boundaries were related to creating or returning to rhetorical goals.
>
> *Prediction 4:* Key acts of rhetorical planning were setting, reviewing, and evaluating content and process goals and acting on those goals.

They concluded that (1) planning goes on at many levels: the sentence level and the rhetorical level of audience, task, and writer's own goals; (2) the composing process has an episodic pattern not dictated by the pattern of the text; (3) the beginnings of episodes are dominated by goal-setting activities—one of the chief outputs of the writer's pregnant pause (pp. 239–243). Their original hypotheses helped shape these conclusions. They related their findings to the research of others, thus helping to increase their ability to generalize from their study. Because, however, they stated their conclusions in terms of expert or novice writers in general, they may inadvertently have encouraged readers to extend some of their findings beyond the data.

Case study research has a number of advantages and some disadvantages. These are discussed at the end of the next chapter on ethnography, the second kind of qualitative descriptive research we will examine.

Summary

The case study is a type of qualitative descriptive research that closely examines a small number of subjects, and is guided by some theory of writing. Researchers collect data using a variety of methods such as observations, interviews, protocols, tests, examination of records, and collection of writing samples. They analyze the data, identifying, operationally defining, and relating variables, and then report results in the form of extensive descriptions, conclusions, hypotheses, and questions for further research. Below are a few questions that can guide the reading of case studies.

Checklist for reading "Case Studies"

1. What theory, questions, or hypotheses govern the research?
2. Who are the subjects and why were they chosen?
3. By what means were the data collected? What data were gathered?
4. What variables were identified and operationally defined?
5. In what ways were the variables interrelated, tabulated?
6. What level of agreement was reached by the coders? How was it calculated?
7. In what format were the results reported? Summaries? Figures? Tables? Extensive description?
8. What problems, if any, afflict the study?

References

Research Design and References

Asher, J.W. (1976). *Educational research and evaluation methods.* Boston: Little, Brown.

Atlas, M. (1980). *A brief overview of research methods for the writing researcher.* Document Design Project Working Paper. Carnegie-Mellon University, Pittsburgh, Pa.

Bond, S., & Hayes, J.R. (1980). *Practical aspects of collecting a thinking-aloud protocol.* Working paper, Carnegie-Mellon University, Pittsburgh, Pa.

Braddock, R., Lloyd-Jones, R., & Schoer, L. (1963). *Research in written composition.* Urbana, Ill.: National Council of Teachers of English.

Bridwell, L.S., Nancarrow, P.R. & Ross, D. (1984). The writing process and the writing machine: Current research on word processors relevant to the teaching of composition. In R. Beach, & L.S. Bridwell (Eds.). *New directions in composition research.* New York: Guilford Press.

Cohen, J. (1960). A coefficient of agreement for nominal scales. *Educational and Psychological Measurement, 20,* 37–46.

Ericsson, K.A. & Simon, H.A. (1980). Verbal reports as data. *Psychological Review, 87,* 215–251.

Ericsson, K.A., & Simon, H.A. (1984). *Protocol analysis, verbal reports as data.* Cambridge, Mass.: MIT Press.

Flower, L., & Hayes, J. (1980). Identifying the organization of writing processes. In Lee W. Gregg and Erwin R. Steinberg (Eds.). *Cognitive processes in writing.* Hillsdale, N.J.: Lawrence Erlbaum.

Flower, L., & Hayes, J. (1981a). The pregnant pause: An inquiry into the nature of planning. *RTE, 15,* 229–243.

Flower, L., & Hayes, J. (1981b). A cognitive process theory of writing, *CCC, 32,* 365–387.

Frase, L.T. (1980). *Writer's Workbench: Computer supports for components of the writing process.* Technical Report. Murray Hills, N.J.: Bell Laboratories.

Hayes, J., & Flower, L.S. (1983). Uncovering cognitive processes in writing: An introduction to protocol analysis. In P. Mosenthal, L. Tamor, and S. Walmsley (Eds.). *Research on writing: Principles and methods.* New York: Longman.

Jacob. E. Qualitative research traditions: A review. *Review of Educational Research, 57,* 1–50.

Kerlinger, F.N. (1986). *Foundations of behavioral research,* 3rd edition. New York: Holt, Rinehart & Winston.

Kirk, J. & Miller, M. (1985). *Reliability and validity in qualitative research.* London: Sage Publications.

Krippendorff, K. (1980). *Content analysis: An introduction to its methodology.* London: Sage Publications.

Miles, M., & Huberman, M. (1984). *Qualitative data analysis.* London: Sage Publications.

Milic, L.T. (1981). Stylistic + computers = pattern stylistics. *Perspectives in Computing, 1,* 4–11.

Newell, A., & Simon, H.A. (1972). *Human problem solving.* Englewood Cliffs, N.J.: Prentice-Hall.

Nisbett, R.E., & Wilson, T.D. (1977). Telling more than we can know: Verbal reports on mental processes. *Psychological Review, 84*(3), 231–259.

Nunnally, J.C. (1978). *Psychometric theory,* 2nd edition. New York: McGraw-Hill.

Odell, L. (1974). Measuring the effect of instruction in pre-writing. *RTE, 24,* 228–240.

Perkins, D.N. (1981). *The mind's best work.* Cambridge, Mass.: Harvard Univ. Press.

Perl, S. (1984). *Coding the composing process: A guide for teachers and researchers.* ERIC Document Reproduction Service No. ED 240 609.

Schwartz, H.J., & Bridwell, L.S. (1984). A selected bibliography of computers in composition. *CCC, 35,* 71–77.

Simon, H.A. (1979). *Models of thought.* New Haven, Conn.: Yale Univ. Press.

Simons, H. (ed.) (1980). *Towards a science of the singular: Essays about case study in educational research and evaluation.* CARE Occasional Paper No. 10, University of East Anglia.

Smith, J.K., & Heshusius, L. (1986). Closing down the conversation: The end of the quantitative–qualitative debate among educational inquirers. *Educational Researcher, 15*(1), 4–12.

Stone, P.J., et al. (1966). *The General Inquirer: A computer approach to content analysis.* Cambridge, Mass.: MIT Press.

Van Maanen, J. (1983). *Qualitative methodology.* London: Sage Publications.

Vockell, E.L. (1983). *Educational research.* New York: Macmillan.

Witte, S., and Davis, A. (1982). The stability of the T-unit length in the written discourse of college freshmen: A second study. *RTE, 16,* 71–84.

Yin, R.K. (1984). *Case study research.* London: Sage Publications.

Young, R.E., Becker, A.P., & Pike, K. (1970). *Rhetoric: Discovery and change.* New York: Harcourt Brace Jovanovich.

Case Studies

Balkema, S.B. (1984). The composing activities of computer literate writers. *DAI, 45,* 12A.

Bechtel, J. (1979). The composing processes of six male college freshman enrolled in technical programs. *DAI, 39,* 09A.

Birnbaum, J. (1982). The reading and composing behavior of selected fourth- and seventh-grade students. *RTE, 16,* 241–260.

Bissex, G. (1980). Patterns of development in writing—a case study. *Theory into practice, 19,* 197–201.

Bond, S., Hayes, J., & Flower, J. (1980). *Translating law into common language: A protocol study.* Technical Report No. 8. Pittsburgh, Pa.: Carnegie-Mellon University.

Booley, H.A. (1984). Discovery and change: How children redraft their narrative writing. *Educational Review, 36*(3), 263–275.

Bryant, D.G. (1984). The composing processes of blind writers. *DAI, 45,* 11A.

Butler, M. (1983). Levels of engagement, rhetorical choices, and patterns of differentiation in the writing of four eleventh graders. *DAI, 45,* 02A.

Calkins, L. (1980). Children's rewriting strategies. *RTE, 14,* 331–341.

Cayer, R.L., & Sacks, R. (1979). Oral and written discourse of basic writers: Similarities and differences. *RTE, 13,* 121–128.

Cooper, C., & Odell, L. (1976). Considerations of sound in the composing process of published writers. *RTE, 10,* 103–115.

Drexler, N.G. (1984). Conceptions and practices of writing among six- to ten-year-old children: Fiction, non-fiction, and inner speech. *DAI, 45,* 06A.

Dyson, A.H. (1984a). Emerging alphabetic literacy in school contexts. *Written Communication, 1,* 5–55.

Dyson, A.H. (1984b). Learning to write/Learning to do school: Emergent writers' interpretations of school literacy tasks. *RTE, 18*(3), 233–264.

Emig, J. (1971). *The composing processes of twelfth graders.* Urbana, Ill.: National Council of Teachers of English.

Farr, M., & Janda, M.A. (1985). Basic writing students: Investigating oral and written language. *RTE, 19*(1), 62–83.

Flaherty, G.P. (1984). The speaking/reading/writing connection interaction in a basic reading classroom. *DAI, 45,* 08A.

Flower, L., & Hayes, J. (1980). The cognition of discovery: Defining a rhetorical problem. *CCC, 31,* 21–32.

Flower, L., Hayes, J., & Swarts, H. (1984). Reader-based revision of functional documents: The scenario principle. In P. Anderson, C. Miller, & J. Brockmann (Eds.). *New essays in technical and scientific communication: Theory, research, and criticism.* Farmington, N.Y.: Baywood Series in Technical and Scientific Communication, Vol. 2.

Fontaine, S.I. (1984). Writing for an audience: How writers at three age levels demonstrate an awareness of the audience and respond to two contrasting audiences. *DAI, 45,* 06A.

Gere, A. (1982). Insights from the blind: Composing with revising. In Ronald Sudol (Ed.). *Revising* Urbana, Ill.: National Council of Teachers of English.

Glassner, B. (1980). Preliminary report: Hemispheric relationships in composing. *Journal of Education, 162,* 74–95.

Graves, D. (1973). Children's writing: Research directions and hypotheses based upon an examination of the writing process of seven-year-old children. *DAI, 34,* 6255A.

Graves, D. (1975). An examination of the writing processes of seven-year-old children. *RTE, 9,* 227–241.

Graves, D. (1979). What children show us about revision. *Language Arts, 56,* 312–319.

Hamer, J.A. (1984). Describing revision: How selected ninth and twelfth grade students revise narratives. *DAI, 45,* 03A.

Healy, M.K. (1984). Writing in a science class: A case study of the connections between writing and learning. *DAI, 45,* 07A.

Herrington, A.J. (1983). Writing in academic settings: A study of the rhetorical contexts for writing in two college chemical engineering courses. *DAI, 45,* 01A.

Hilgers, T. (1984). Toward a taxonomy of beginning writers' evaluative statements on written compositions. *Written Communication, 1*(3), 365–384.

Holland, N. (1982). Theories can be applied: A study of a theory-based writing class. *Clearing House, 55,* 248–252.

Hudson, S.A. (1984). Contextual factors and children's writing. *DAI, 45,* 06A.

Jolliffe, D.A. (1984). Audience, subjects, form and lexicon: Writers' knowledge in three disciplines. *DAI, 46,* 367A–368A.

Jones, N. (1982). Design, discovery, and development: A case study. *DAI, 43,* 04A.

Kasten, W.C. (1984). The behaviors accompanying the writing process in selected third- and fourth-grade native American children. *DAI, 45,* 08A.

Katz, S. (1984). Teaching the tagmemic discovery procedure: A case study of a writing course. *DAI, 45,* 05A.

Keech, C.L. (1984). Apparent regression in student writing performance as a function of unrecognized changes in task complexity. *DAI, 45,* 09A.

Linn, B. (1978). Psychological variants of success: four in-depth case studies of freshmen in a composition course. *College English, 39,* 903–917.

Matsuhashi, A. (1981). Pausing and planning: The tempo of written discourse production. *RTE, 15,* 113–134.

Mischel, T. (1974). A case study of a twelfth-grade writer. *RTE, 8,* 303–314.

Monohan, B.D. (1984). Revision strategies of basic and competent writers as they write for different audiences. *RTE, 18*(3), 288–304.

Morgan, M. (1988). *The composing processes of student collaborative writers.* Doctoral dissertation, Purdue University.

Nichols, R.G. (1984). The effects of computer-assisted writing on the composing processes of basic writers. *DAI, 45,* 08A.

Nystrand, M. (1979). Using readability research to investigate writing. *RTE, 13,* 231–242.

Odell, L., & Cooper, C. (1976). Describing responses to works of fiction. *RTE, 10,* 203–225.

Odell, L., Goswami, D., & Herrington, A. (1983). The discourse-based interview: A procedure for exploring the tacit knowledge of writers in nonacademic settings. In P. Mosenthal, L. Tamor, and S. Walmsley (Eds.). *Research on writing: Principles and methods.* New York: Longman.

Onore, C.S. (1983). Students' revisions and teachers' comments: Toward a transactional theory of the composing process. *DAI, 45,* 06A.

Perl, S. (1979a). The composing process of unskilled college writers. *RTE, 13,* 317–336.

Perl, S. (1979b). Five writers writing: Case studies of the composing processes of unskilled college writers. *DAI, 39,* 08A.

Pollard, R.H. (1984). Fourth graders' personal narrative writing: A study of perceptions of personal narrative discourse and of narrative composing decisions and strategies. *DAI, 45,* 02A.

Reagan, S.B. (1984). The effect of combined reading–writing instruction on the composing processes of basic writers: A descriptive study. ERIC Document Reproduction Service No. ED 243 134.

Rental, V., & King, M. (1983). Present at the beginning. In P. Mosenthal, L. Tamor, and S. Walmsley. *Research on writing: Principles and methods.* New York: Longman.

Rubin, L.E. (1984). How student writers judge their own writing. *DAI, 45,* 03A.

Rule, R. (1982). Research update: The spelling process—a look at strategies. *Language Arts, 59,* 379–384.

Schrader, C.T. (1984). The influence of oral language transactions between selected teachers and prekindergarten children on developing literacy during early written language. *DAI, 45,* 11A.

Selfe, C.L. (1984). The predrafting processes of four high- and four low-apprehensive writers. *RTE, 18,* 45–64.

Selfe, C.L. (1984b). *Reading as a writing and revising strategy.* ERIC Document Reproduction Service No. ED 244 295.

Shanahan, J.B. (1984). Reader response literature: A source for effective descriptors of the revision stage in the process of writing. *DAI, 45,* 08A.

Shock, D.H. (1984). The writing process: Effects on life-span development on imaging. *DAI, 45,* 06A.

Smith, J.P. (1982). Writing in a remedial reading program: A case study. *Language Arts, 59,* 245–253.

Sommers, N. (1979). Revision in the composing process: A case study of college freshman and experienced writers. *DAI, 45,* 09A.

Sommers, N. (1980). Revision strategies of student writers and experienced writers. *CCC, 31,* 378–388.

Sowers, S. (1979). A six-year-old's writing process—the first half of first grade. *Language Arts, 56,* 829–835.

Stallard, C. (1974). An analysis of the writing behavior of good student writers. *RTE, 8,* 206–218.

Strickland, R.W. (1984). A case study examination of reader awareness and the composing process of undergraduate business students. *DAI, 45,* 05A

Thomas, D.K. (1984). A transition from speaking to writing: Small-group writing conferences. *DAI, 45,* 09A.

Thompson, E.H. (1984). The effect of feedback system on teacher performance in writing conferences. *DAI, 45,* 07A.

Wagner, M.J. (1984). A comparison of fifth graders' oral and written stories. *DAI, 45,* 08A.

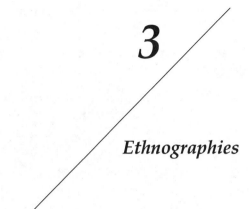

3

Ethnographies

Ethnographic research, another kind of qualitative descriptive research, examines entire environments, looking at subjects in context. This design derives primarily from phenomenology, anthropology, and sociology, which have argued for its importance as a research method to provide a window on culture. In contrast to case studies, ethnographers observe many facets of writers in their writing environments over long periods of time in order to identify, operationally define, and interrelate variables of the writing act in its context. The assumption behind this design is that to understand human behavior which occurs in a natural setting, one must view it as a part of that environment. The researcher often becomes a *participant observer*, a member of the classroom or other situations being studied, with a minimum of overt intervention. Based on extensive observations, researchers generate hypotheses, validate them by returning to the data, and produce *thick descriptions*, detailed accounts of writing behavior in its rich context.

A number of ethnographies have been done on various dimensions of composition. Nelson and Kantor review work in three areas: developmental processes, the classroom community, and writing instruction (1981). Kantor, Kirby, and Goetz also discuss ethnographic studies in several categories of English education: language, composition, reading, and literature (1981). We will use two studies to examine this design.

Two ethnographies

Study 1   Susan Florio and Christopher Clark (1982). The functions of writing in an elementary classroom. *RTE, 16,* 115–130.

Florio and Clark studied a second- and third-grade classroom where they observed the role of writing in the lives of the children. We will discuss this study below.

Study 2   Alan Lemke and Lillian Bridwell (1982). Assessing writing ability—an ethnographic study of consultant–teacher relationships. *English Education, 14,* 86–94.

Lemke and Bridwell studied the process of evaluating an elementary writing program for a school district. They interacted with teachers and administrators in this environment in an effort to develop criteria and to assess student progress.

Research design: ethnography

Research process

Overall, researchers identify and define the whole environment they plan to study. They then plan ways of varying their observations and gaining multiple perspectives by mapping the setting, selecting observers and developing a relationship with them, and establishing a long period of investigation. This multiplicity of observations is called *triangulation.* Undergirding triangulation is a conception of knowledge as a social construction, a collaborative search, interpretation, and reinterpretation of complex acts in context.

The validity of this kind of research comes from a continual reciprocity between developing hypotheses about the nature and patterns of the environment under study and the regrounding of these hypotheses in repeated observation, further interviews, and the search for disconfirming evidence. The next sections will examine this process in more detail.

Selection of environment

In selecting an environment to study, a researcher chooses one that promises to yield hypotheses about the behavior the researcher wants to understand.

Hymes (1980) states that the validity of an ethnography depends on an "accurate knowledge of the meanings and behaviors of those who participate in them (p. 93)." The choice of site and sample selection is important because it determines the extent to which the researchers' conclusions about any aspect of writing or writing instruction will be considered representative of other environments. A researcher who studies a unique or unusual environment limits the extent to which the conclusions will be deemed significant in the field and worth further investigation. Another consideration in selecting an environment is its availability to the researcher.

Florio and Clark were prompted to do an ethnography because the National Assessment of Educational Progress reported in 1981 that while children's writing improved, their enjoyment and sense of competence declined. To ascertain why this condition existed, they undertook to study a second- and third-grade open classroom located in a midwestern city with a large land grant university, a site convenient for the researchers. The classroom housed two second–third grades and two fourth–fifth grades with four teachers. The room had movable chairs and a common area. The students represented a mixture of parental economic status (university, state employees, working class) and diverse racial and ethnic backgrounds. A schoolwide K–12 curriculum revision gave high priority to writing instruction, provoking a range of attitudes on the part of teachers and parents. Because the researchers wanted to study attitudes toward writing, this classroom seemed to offer a good environment, even though it was somewhat atypical, a characteristic they acknowledged and discussed in relation to their goals.

Lemke and Bridwell also did an ethnographic study on their process of evaluating an elementary writing program in a suburban school district. They studied the process by which they, as experts, shared their technical knowledge with members of the district as a way of involving the teachers in directing their own evaluation—making the major decisions and writing the final report. They were guided by theories from research on writing evaluation, experience with change processes in schools, and familiarity with ethnographic research. Their research is a good example of how an ethnographic study can be carried out in conjunction with responsibilities the researcher already has.

Role of the Ethnographer

There are at least two kinds of roles that the researcher can play. In one case, the researcher can act as observer outside the scene. In classrooms, students and teachers generally accustom themselves to the presence of observers and video equipment after about two weeks. The ethnographer can also become a participant observer, acting as an assistant teacher in a classroom, a fellow worker in a professional situation, or an official visitor, roles that are not unusual in many environments. Researchers must be careful to report the types of roles they played.

Also important to discuss will be the perspectives that the researcher brought to the environment, perspectives that are usually discussed in the problem statement and literature review. No ethnographer can be "objective," nor is that the goal. The researcher's perspective becomes an important part of the environment studied.

Data collection: triangulation

As we mentioned above, researchers plan ways of obtaining many perspectives: e.g., using multiple observers, collecting writing samples, conducting interviews with students and instructors, and taking copious notes. Agar (1985) suggests that three kinds of notes be taken: observational (a record of the ongoing activities), methodological (a record of the means of triangulation), and theoretical (speculations about theory as it suggests itself to account for the behaviors under scrutiny).

As participant observers in the class, Florio and Clark took extensive field notes, made videotapes of everyday life in the classroom, gathered weekly journals by the teachers who described their general method of instruction and of teaching writing, and interviewed the teachers about the content of the journals and videotapes. They also collected students' written work and held conferences with students about this work. This combining of multiple sources of data is called *triangulation,* an important feature of good ethnographic research.

Lemke and Bridwell also used a variety of techniques to collect data. For each session, one acted as participant observer, taking field notes while the other led the discussion. A research assistant took a second set of field notes to provide greater reliabilities and more than one perspective. The sessions included brainstorming discussions, T-unit counts, and a survey of teaching practices.

Data analysis

One of the challenges of ethnographic research is coding the data collected from a variety of sources. Researchers must create categories and notice patterns and relationships among the data. Here again the source of these categories is twofold: the data themselves as analyzed by the researcher and preferably other coders, and categories or variables already identified in previous research and theory. The researcher creates hypotheses about these categories and their relationship based on segments of the data. Agar calls these segments "strips," any bounded phenomenon against which an ethnographer tests the material (1985). The goal is to create and test schemas that will account for and explain the strips of writing behavior in context. Important here is the reliability among coders and the testing of schemas in new environments.

Because of the mixture of student backgrounds, the role that writing played in building community, and the learning choices that the room allowed, Florio and Clark were able to identify the following variables as functions of writing: (1) participating in community, (2) knowing oneself and others, (3) occupying free time, and (4) demonstrating academic competence. In their report, they provided thick descriptions of these functions, including samples of individuals' writing (1982, pp. 120–128).

Lemke and Bridwell fully explained their three-step process of analysis. They and an assistant analyzed the data by responding independently to the two sets of field notes and then met to analyze their independent interpretations. These discussions resulted in "issue summaries" that became the variables characterizing their interactive process of evaluation: (1) the use of technical terms, (2) the role of the consultant (advisor, superior, director, expert), (3) consultant agendas, (4) task adherence and practicality, and (5) the context. They explained and exemplified each of these variables that influenced the evaluative process (pp. 88–93).

Results

A researcher does not have the freedom to observe without restrictions and to report results as ultimate truth. As in any good scholarship, issues of generalizability of observations and data and relationships among variables must be considered.

Both studies published reports that included a good deal of thick description. Florio and Clark presented samples of students' writing, a merit especially in the light of the limited space in journals. See Figure 3-1. As good ethnographers, they also reported conclusions that suggested important variables for further investigation: that writing was a part of the school lives of the children in room 12, involving them in

Student letter

Lab booklet page

Behavior contract

Diary page

Figure 3-1 Documents reflecting the functions of writing in room 12. *Source:* Florio and Clark (1982), Figure 2. Copyright © 1982 by the National Council of Teachers of English. Reprinted by permission of the publisher.

the four functions described above. In more detail they pointed out unsettling results.

1. Despite the range of opportunities to write, those directly concerned with the social and academic aspects of school life relieved the students of important roles and responsibilities for writing.
2. Only one type of writing was evaluated—that done in collaboration.
3. The diary, the one type that involved the student directly in the entire process, was least formal and visible to the people concerned with formal schooling.
4. The most promising function, which started with the real experiences of children, was the most challenging pedagogically.

Overall, they concluded that the organization of the social context of the classroom could either enable and legitimate a range of functions or force much of the writing process underground, leaving for assessment only that kind that entails minimal commitment (1982, pp. 128–129). In all of these conclusions, they were careful not to generalize beyond the subjects of the study.

Lemke and Bridwell reported several conclusions: (1) the use of technical terms disturbed the relationship between the consultants and the teachers, (2) the role of consultant was difficult to maintain and needed to be modified to establish symbiosis, (3) they were often oppressive in their attempt to influence teachers to adopt views about writing and evaluation consistent with their own philosophy, (4) practical values dominated the teachers' attitudes toward research conclusions, and (5) the teachers' knowledge of context (school board, district teachers, and school administration) was one basis for their evaluating the data produced. They generally concluded that a researcher's challenge is to develop knowledge in context and a teacher's is to demand it (pp. 93–94).

Advantages and disadvantages of case study and ethnography

Advantages

Qualitative research attempts to give a rich account of the complexity of writing behavior, a complexity that controlled experiments generally cannot capture. It tries to show the interrelationships among multifaceted dimensions of the writing process by looking closely at writing from a new point of view in order to recognize important variables and to suggest new hypotheses for further study. It is often the most feasible and useful kind of research for composition instructors because it is closest to the teacher's experience and can be done in the classroom. Its results can be exceptionally helpful in training teachers, because they can begin to understand the full complexity of the composition classroom.

Cook and Campbell (1979) point out another benefit: Qualitative holistic, contextual research is at its best when it exposes the blindness and gullibility of specious quantitative research. They strongly urge that "field experimentation should always include qualitative research to describe and illuminate the context and conditions under which research is conducted" (p. 93).

Difficulties and problems

Case study and ethnography also have difficulties, pitfalls, and cautions. Cook and Campbell (1979) make this point clearly when they discuss problems with the ability of studies to make cause-and-effect conclusions and to generalize, problems they contend are "not limited to quantitative or deliberately experimental studies. They arise in *less formal, more commonsense, humanistic, "global," contextual integrative and qualitative* approaches to knowledge" (p. 92).

Sadler, in an article on cognitive limitations in qualitative data from holistic naturalistic settings, explains that while the human brain is unmatched in its ability to derive information from noisy environments, empirical studies have shown that it can also derive recurring patterns of fallible and incorrect inferences from data. Naturalistic qualitative data processing is capable of misinterpretation, misaggregation, and defective inferences (1981).

Sadler lists several problems of interpreting qualitative data. The first is *data overload:* it is well known that people have severe limitations on the amount of data they can receive, remember, and process, especially when inferences are to be made. Data may be so extensive as to inhibit adequate analysis. A second problem is *first impressions:* the order in which the information is received may dominate the researcher's judgment. A related pitfall is *confidence in judgment.* It has been shown empirically that people have such a profound confidence in the accuracy of their decisions once made, that they will not change even in the light of considerable, relevant, and contrary evidence. A third problem, *availability of information,* rests on the fact that the plausibility of an interpretation depends on the ease with which the researcher and the reader of the research are able to retrieve concrete instances or examples of the phenomenon, and the ease with which the reader can understand the reasoning in the reporting. The opposite effect can also occur when a researcher relies more heavily on information readily obtained, for example, orally through personal contacts or via anecdotes or graphs, even when far more comprehensive and complete information is available. A fourth problem is *positive or negative instances*. Sadler states: "When tentative hypotheses are held, . . . evidence is unconsciously selected in such a way that it tends to confirm the hypotheses. In other words, what is noticed, or what counts as a fact, depends in part on what is to be verified" (1981, p. 28). Even careful, intelligent people tend to adhere to their original beliefs; even when conflicting evidence is available, they simply do not recognize it.

Sadler's fifth concern is a cluster composed of *internal consistency, redundancy*, and *novelty of information*. It has been shown that there is a cognitive tendency to overweigh the importance of more extreme or novel data. On the other hand, redundancy of information tends to reduce its perceived importance. Sadler notes "that balanced conflicting information does not 'cancel out' and leave one with no knowledge at all. To be aware of ambiguity is not the same as being entirely ignorant" (p. 28). A sixth problem is *uneven reliability of information*. Data from poor sources are treated about the same as reliable data. While people do tend ultimately to discount data from less than credible sources, nevertheless, the revisions of earlier held hypotheses are not always sufficient to overcome first impressions. *Missing information* is a seventh problem for qualitative data interpretation. Some researchers tend to fill in missing data when evaluating it, often in unpredictable ways—some overrating it, some underrating it, and some rating it as average. The type of rating is affected by what researchers consider important conceptually and also by the focus of their attention at the time of the operational development of its interpretation. Finally, when comparative data are more or less available on several variables, their absence or presence may result in a stronger emphasis on a particular variable in making decisions about people or programs.

An eighth problem is *base-rate proportion*. Researchers are often insensitive to the total size of a population, basing impressions on only a few cases, or even a single exceptional case. A ninth difficulty stems from *sampling considerations*. It is well known from sampling theory that small samples exhibit less reliability of data than large samples. Yet even experienced users of statistics expect less variation than statistical theory dictates. People generally are less sensitive to sample size than to the representativeness of the population. They expect population characteristics to hold true in the sample and local results to hold true for the population. In case studies and ethnography, usually no rigorous sampling procedure has been applied. The selection of subjects is based, as we indicated, on the usefulness of the subjects for studying what the investigator wants to analyze and on the availability of the subjects.

A tenth problem in naturalistic settings is *co-occurrences and correlation*. It has been shown that observed co-occurrences are often misinterpreted as cause-and-effect relationships. Smedslund (1963), for instance, has found that "adults with no training in statistics . . . typically reveal . . . an exclusive dependence on + +[double positive] instances" (p. 172). In other words, they tend to interpret observed co-occurences as evidence of strong correlation. An example from a hypothetical ethnographic study on writing would be the conclusion that the use of journals in a particular classroom was the *cause* of an increase in syntactic fluency, when, in fact, these two variables had only been observed as co-occurring.

In addition to the problems with representativeness, cause-and-effect judgments, and variable development identified above, other difficulties attend qualitative research. It is time-consuming, takes considerable writing skill, and requires more journal space to publish results. Researchers must also be concerned about its replicability, its repeatability with the same results. Will the same variables be gleaned from new settings by other observers? Will these variables remain stable over time? Qualitative research has a limited ability to generalize to other samples, variables, and conditions like the ones studied. In composition research, sometimes readers who do not have strong empirical research backgrounds interpret the results of qualitative research as definitive rather than as exploratory and generative of hypotheses.

Summary

Ethnography engages researchers, sometimes as participant-observers, in a study of writing in context. Though guided by theory and questions or hypotheses, investigators withhold initial judgments, allowing the data to determine particular research directions. Using a variety of methods, they collect a rich array of data, taking field notes, interviewing, collecting writing samples and whatever other information is available. They analyze and code the data, identifying, defining, and relating what seem to be important variables, and finally report their study in the form of thick descriptions.

Checklist for reading "Ethnographies"

1. What theory, questions, or hypotheses govern the research?
2. Who are the subjects and environments and why were they chosen?
3. By what means were the data collected? What data were gathered?
4. What variables were identified and operationally defined?
5. What role did the researcher(s) play?
6. In what ways were the variables interrelated?
7. In what form or detail was the study reported?
8. What conclusions were drawn?
9. What problems, if any, afflict the study?

References

Research Design and References

Agar, M.H. (1985). *Speaking of ethnography.* London: Sage Publications.

Agar, M.H. (1982). Toward an ethnographic language. *American Anthropologist, 84,* 779–795.

Bogdan, R., & Taylor, S.J. (1975). *Introduction to qualitative research methods: A phenomenological approach to the social sciences.* New York: Wiley.

Brodkey, L. (1987). Writing ethnographic narratives. *Written Communication, 9,* 25–50.

Burgess, T. (1984). *Language in classrooms: A theoretical examination of different research perspectives and an outline of a socio-cultural approach to classroom discourse.* Unpublished Ph.D. dissertation, University of London Institute of Education.

Clark, C.M., & Florio, S. (1983). Understanding writing in school: Issues of theory and method. In P. Mosenthal, L. Tamor, & S. Walmsley (Eds.). *Research on writing: Principles and methods.* New York: Longman.

Cole, M., Hood, L., & McDermott, R. (n.d.). *Ecological niche picking: Ecological validity as an axiom of experimental cognitive psychology.* Laboratory of Comparative Human Cognition and Institute for Comparative Human Development, New York: The Rockefeller University.

Cook, T.D., & Campbell, D.T. (1979). *Quasi-experimentation: Design and analysis issues for field settings.* Boston: Houghton Mifflin.

Doheny-Farina, S., & Odell, L. (1986). Ethnographic research on writing: Assumptions and methodology. In L. Odell & D. Goswami (Eds.). *Writing in non-academic settings.* New York: Guilford Press.

Douglas, J.D. (1976). *Investigative social research: Individual and team field research.* Beverly Hills, Calif.: Sage Publications.

Emerson, R.M. (Ed.) (1983). *Contemporary field research.* Boston: Little, Brown.

Erickson, F. (1973). What makes school ethnography ethnographic. *Anthropology and Education Quarterly, 4*(2), 10–19.

Fetterman, D. (1982). Ethnography in educational research: The dynamics of diffusion. *Educational Researcher, 11,* 17–22.

Garfinkel, H. (1967). *Studies in ethnomethodology.* Englewood Cliffs, N.J.: Prentice-Hall.

Geertz, C. (Ed.) (1973a). *The interpretation of culture.* New York: Basic Books.

Geertz, C. (1973b). Thick description: Toward an interpretative theory of culture. In C. Geertz (Ed.). *The interpretation of culture.* New York: Basic Books.

Georges, R.A., & Jones, M.O. (1980). *People studying people: The human element in fieldwork.* Berkeley: Univ. of California Press.

Glaser, B. (1969). The constant comparative method of qualitative analysis. In McCall, M.J. & Simmons, J.L. (Eds.). *Issues in participant observation.* Reading, Mass.: Addison–Wesley, pp. 216–228.

Goetz, J.P., & Le Compte, M.D. (1984). *Ethnography and qualitative design in educational research.* Orlando, Fla.: Academic Press.

Green, J.L., & Wallot, C. (Ed.) (1981). *Ethnography and language in educational settings,* Vol. 5, *Advances in discourse processes.* Norwood, N.J.: Ablex.

Guba, E. (1980). *Naturalistic and conventional inquiry.* Paper presented at the annual meeting of American Educational Research Association, Boston, April.

Hammersley, M., & Atkinson, P. (1983). *Ethnography.* London: Tavistock.

Hymes, D.H. (1964). Introduction: Toward ethnographies of communication. *American Anthropologist, 66,* 1–34.

Hymes, D.H. (1972). Introduction: Toward ethnographies of communication. In P. Giglioli (Ed.). *Language and social context.* Baltimore, Md.: Penguin.

Hymes, D.H. (1980). What is ethnography? In D. Hymes (Ed.). *Language in education: Ethnographic essays.* Arlington, Va.: Center for Applied Linguistics.

Kantor, K. (1984). Classroom contexts and the development of writing intuitions: An ethnographic case study. In R. Beach and L.S. Bridwell (Eds.). *New directions in composition research.* New York: Guilford Press.

Kantor, K., Kirby, D.R., & Goetz, J.P. (1981). Research in context: Ethnographic studies in English education. *RTE, 15,* 293–309.

Le Compte, M.D., & Goetz, J.P. Problems of reliability and validity in ethnographic research. *Review of Educational Research, 52,* 31–60.

Lincoln, Y.S., & Guba, E.G. (1985). *Naturalistic inquiry.* London: Sage Publications.

Marcus, G.E., & Cushman, D. (1982). Ethnographies as texts. *Annual Review of Anthropology, 11,* 25–69.

Martin, J. (1982). A garbage can model of the research process. In J.E. McGrath, J. Martin, & R.A. Kulka (Eds.). *Judgment calls in research.* Beverly Hills, Calif.: Sage Publications.

McGrath, J. (1982a). Dilemmatics: The study of research choices and dilemmas. In J.E. McGrath, J. Martin, & R.A. Kulka (Eds.). *Judgment calls in research.* Beverly Hills, Calif.: Sage Publications.

McGrath, J. (1982b). Idiosyncrasy and circumstance: Choices and constraints in the research process. In J.E. McGrath, J. Martin, & R.A. Kulka (Eds.). *Judgment calls in research.* Beverly Hills, Calif.: Sage Publications.

Merton, R. (1972). Insiders and outsiders: A chapter in the sociology of knowledge. *American Journal of Sociology, 78,* 9–47.

Miles, M., & Huberman, A.M. (1984). *Analyzing qualitative data.* Beverly Hills, Calif.: Sage Publications.

Mishler, E. (1979). Meaning in context: Is there any other kind? *Harvard Educational Review, 49,* 1–19.

Morgan, G. (Ed.) (1983) *Beyond method: Strategies for social research.* Beverly Hills, Calif., & London: Sage Publications.

Nelson, M., & Kantor, K. (1981). Context-dependent studies in composition: A selective review. *Research in Composition Newsletter*, 2, 1–7.

Overholt, G. (1980). Ethnography and education: Limitations and sources of error. In A.A. Van Fleet (Ed.). *Anthopology of education: Methods and applications*. [Special Issue] *Journal of Thought*, 15, 1–20.

Pellegrini, A.D., & Yawley, T.D. (Eds.) (1985). *The development of oral and written language in social contexts*. Norwood, N.J.: Ablex.

Punch, M. (1986). *The politics and ethics of fieldwork*. Beverly Hills, Calif.: Sage Publications.

Sadler, D.R. (1981) Intuitive data processing as a potential source of bias in naturalistic evaluations. *Educational Evaluation and Policy Analysis*, 3(4), 25–31.

Schatzman, L., & Strauss, A.L. (1973). *Field research: Strategies for a natural sociology*. Englewood Cliffs, N.J.: Prentice-Hall.

Smedslund, J. (1963). The concept of correlation in adults. *Scandinavian Journal of Psychology*, 4, 165–173.

Spindler, G. (Ed.) (1982). *Doing the ethnography of schooling*. New York: CBS College Publishing.

Szwed, J.F. (1981). The ethnography of literacy. In M. Farr Whiteman (1980) (Ed.). *Writing: The nature, development, and teaching of written communication*. Hillsdale, N.J.: Laurence Erlbaum, p. 13–23, 72–94.

Werner, O., & Schoepfle, G.M. (1986). *Systematic fieldwork* (Vols. 1–2). Beverly Hills, Calif.: Sage Publications.

Wilson, S. (1977). The use of ethnographic techniques in educational research. *Review of Educational Research*, 47, 249–265.

Ethnographies

Barabas, C.P. (1984). The nature of information in technical progress reports: An analysis of writer intentions, texts, and reader expectations. *DAI*, 45, 11A.

Basso, K.H. (1974). The ethnography of writing. In R. Bauman & J. Sherzer (Eds.). *Explorations in the ethnography of speaking*. London: Cambridge University Press.

Black, J. (1979). Formal and informal means of assessing the communicative competence of kindergarten children. *RTE*, 13, 49–68.

Blazer, B. (1984). The development of writing in kindergarten: Speaking and writing relationships. *DAI*, 45, 05A.

Braig, D.E. (1984). Six authors in search of an audience: Dialogue journal writing of second graders. *DAI*, 45, 05A.

Britton, J., Burgess, T., Martin, N., McLeod, A., & Rosen, H. (1975). *The development of writing abilities, 11–18*. London: Macmillan & Co.

Clark, C.M., & Florio, S., with Elmore, J.L., Martin, J. Maxwell, R.J. & Metheny, W (1981). *Understanding writing in school: A descriptive study of writing and its instruction in two classrooms*. Final report of the Written Literacy Study (Grant No. 908040) funded by the National Institute of Education, U.S. Department of Education, East Lansing, Mich. Institute for Research in Teaching, Michigan State University. Also in P. Mosenthal, L. Tamor, & S. Walmsley (Eds.). (1983). *Research on writing: Principles and methods*. New York: Longman.

Cowley, K. (1984). Planning and producing writing in the workplace. *DAI*, 45, 07A.

Curtiss, D.H. (1984). The experience of composition and word processing: An ethnographic, phenomenological study of high school seniors. *DAI*, 45, 04A.

Doheny-Farina, S. (1984). Writing in an emergent business organization: An ethnographic study. *DAI, 45,* 11A.

Dyson, A.H. (1984). Learning to write/learning to do school: Emergent writers' interpretations of school literacy tasks. *RTE, 18*(3). 233–264.

Dyson, A.H. (1985). Second graders sharing writing. *Written Communication, 2*(2), 189–215.

Ehrlich, D.B. (1984). A study of the word processor and composing changes in attitude and revision practices of inexperienced student writers in a college composition class. *DAI, 45,* 07A.

Florio, S., & Clark, C. (1982). The functions of writing in an elementary classroom. *RTE, 16,* 115–130.

Freedman, S.W., & Calfee, R.C. (1984). Understanding and comprehending. *Written Communication, 1*(4). 459–490.

Heath, S.B. (1983). *Ways with words: Language, life, and work in communities and classrooms.* London/New York: Cambridge Univ. Press.

Heath, S.B. (1980). What no bedtime story means: Narrative skills at home and school. *Language and Society, 2,* 49–76.

Heath, S.B. (1980). The functions and uses of literacy. *Journal of Communication, 30,* 123–133.

Hymes, D. (1981). Ethnographic monitoring of children's acquisition of reading and language arts skills in and out of the classroom. Vols. 1, 2, 3 (Eric No. Ed 208 096).

Kamler, B. (1980). One child, one teacher, one classroom: The story of one piece of writing. *Language Arts, 57,* 680–693.

Kuhn, M.S. (1984). A discourse analysis of discussions in the college classroom. *DAI, 45,* 06A.

Lemke, A., & Bridwell, L. (1982). Assessing writing ability—an ethnographic study of consultant–teacher relationships. *English Education, 14,* 86–94.

Martin, J.M. (1984). Curriculum of middle school: A descriptive study of the teaching of writing. *DAI, 45,* 08A.

Moxley, J.J. (1984). Five writers' perceptions: An ethnographic study of the composing processes and writing functions. *DAI, 45,* 06A.

Murray, D.E., & IBM Los Angeles Scientific Center. (1985). *Computer conversation: Adapting the composing process to conversation.* Stanford Univ. Press, Stanford, Calif.

Nelson, M.W. (1981). Writers who teach: A naturalistic investigation. *DAI, 42,* 08A.

Odell, L. (1986). Beyond text. In L. Odell & D. Goswami (Eds.). *Writing in nonacademic settings* New York: Guilford Press, pp. 249–278.

Odell, L., & Goswami, D. (Eds.). (1986). *Writing in nonacademic settings.* New York: Guilford Press.

Peterson, B.T. (1984). *Technical writing, revision, and language communities.* ERIC Document Reproduction Service No. ED 245 246.

Pettigrew, J., Shaw, R.A., & Van Nostrand, A.D. (1981). Collaborative analysis of writing instruction, *RTE, 15,* 329–342.

Pipman, M.H. (1984). The amount and nature of composition instruction in two secondary English classrooms. *DAI, 45,* 02A.

Sowers, S. (1979). A six year-old's writing process: The first half of the first grade. *Language Arts, 56,* 829–835.

Wieler, S.H. (1986). *A context-based study of the writing of eighteen year olds, with special reference to A-level biology, English, geography, history, history of art, and sociology.* Unpublished thesis, University of London Institute of Education.

Williamson, M.M. (1984). The function of writing in three college undergraduate curricula. *DAI, 45,* 03A.

Woods-Elliott C.A. (1981). Students, teachers and writing: An ethnography of interactions in literacy. *DAI, 42,* 2376A.

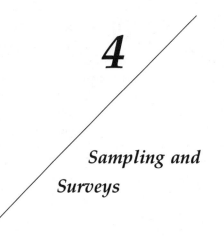

4

Sampling and Surveys

Sampling studies are also descriptive research but of a special kind. Case and ethnographic studies examine one or a few individuals or classrooms in considerable depth, studying the interplay among a large number of variables of an individual or a social context. In contrast, sampling survey research describes a large group, a *population*, of people, compositions, English courses, teachers, or classrooms, in terms of a *sample*, a smaller part of that group. This design concentrates on a few variables of a small group, a sample, which has been drawn to represent a larger population. In statistically designed sampling research, whatever the researcher concludes about the characteristics of the sample can then be considered representative of the entire population. The purpose is to make large descriptive tasks possible with a minimum of cost and effort.

Sampling allows composition researchers to engage in large and otherwise rather unmanageable scholarly efforts. If someone, for example, wanted to assess aspects of 5000 compositions, that task could be reduced by obtaining a representative sample of 500 or even of 50, creating a more reasonable research effort. Even if one chooses to reduce the population to a sample of 50, one can later add other representative samples of 25, 50, 100, and so forth, to the original sample to gain an even clearer description of the population. This research design alleviates much of the difficulty of tedious analysis. But to gain a representative sample of the larger population, one must understand some principles of sampling that will be explained below.

Sampling surveys are a valuable research tool, because they enable the investigator to obtain descriptive information about readily observed or recalled behavior of very large populations. This information can be related to several major demographic characteristics such as age, gender, and formal educational accomplishments, as long as these cross-classifications do not become too numerous. When surveys are used in conjunction with random sampling, investigators can achieve representativeness of large populations.

Three studies

Study 1 Stephen Witte, Paul Meyer, Thomas Miller, & Lester Faigley. (1981). *A national survey of college and university writing program directors*. Technical Report No. 2. Fund for the Improvement of Postsecondary Education Grant G008005896. University of Texas, Austin.

Witte and colleagues tackled a difficult and much-needed piece of composition research. They conducted a large study of program evaluation, supported by the Fund for the Improvement of Postsecondary Education. Before developing an evaluative framework, they decided to collect descriptive data on writing programs by conducting a national survey of college and university writing programs to "provide the profession at large with reliable and current information about college and university writing programs" (p. 1). They chose to do a national survey because, although anecdotal and case study evidence already existed to describe freshman writing programs, they were concerned that too often case studies portrayed single examples as typical while ignoring numerous opposing examples.

They constructed a complex survey instrument to elicit extensive in-depth information rather than just multiple-choice responses. Two outside experts were asked to review the questionnaire. In March 1981, they mailed a letter to over 550 writing program directors, explaining the nature of the survey and the estimated two-hour response time for the questionnaire. Of these, 259 directors, 47%, agreed to complete and return the questionnaire. Of these, 49%, 127 directors, eventually did comply, constituting 23% of the originally identified population of writing program administrators. On the basis of information gleaned from the questionnaire, they described such factors as required and elective writing courses, staffing, proficiency exams, exemption practices, freshman textbooks, activities in courses, evaluation methods, faculty development, teacher workload, amount of required writing, and goals in freshman English.

The researchers claimed that they did not do the survey to test hypotheses, because hypothesis testing assumes a level of prior knowledge and theory that did not exist for writing programs. This claim, however, appears too strong in the light of their later acknowledgment of prior valuable surveys of writing programs. They also reported being guided in constructing their questionnaire by the work of previous researchers and related studies. Thus, theory did guide the scope and dimensions of their description of writing programs as well as the range of activities and methods even if they were not testing hypotheses. We will discuss this study in more detail to illustrate features of sampling survey research.

Study 2   Charlene Eblen (1983). Writing across the curriculum: A survey of a university faculty's views and classroom practices. *RTE, 17*, 343–348.

Eblen developed, pretested, and sent a questionnaire on faculty views about student writing to 471 full-time instructional faculty of the five academic divisions at the University of Northern Iowa. She received 266 returns (56.6%): 54 natural science (66.7%), 26 business (60.5%), 53 social and behavioral sciences (60.2%), 61 education (54%), and 68 humanities and fine arts (46.6%). She analyzed her data, getting information about variables such as academic discipline of instructors, class size, faculty perceptions, and classroom practices. She concluded that the faculty as a whole assigned importance first to overall quality of ideas, and then to organization, development, grammatical form, and coherence. She also found that the problems they perceived fell into two clusters: communicative maturation and standards of edited American English. Of these faculty members, 88% required some writing in their classes; 16% required none. She found that the faculty as a whole concurred with the criteria set by the English department for writing competency. Her design will be discussed below.

Study 3   Betty Bamberg (1978). Composition instruction does make a difference: A comparison of the high school preparation of college freshmen in regular and remedial English classes. *RTE, 12*, 47–59.
Betty Bamberg (1981). Composition in the secondary English curriculum: Some current trends and directions for the eighties. *RTE, 15*, 257–266.

Bamberg studied the changes in the secondary English curriculum in California between 1975 and 1979. In 1975 at UCLA, she selected one section of English 1 and one of Subject A from each scheduled hour between 8 A.M. and 2 P.M. and gave a questionnaire about their high-school English to students in these sections. In 1979 she gave the same questionnaire to students enrolled in six sections of beginning and six sections of advanced composition, selected from the scheduled hours between 9 A.M. and 1:00 P.M. After excluding 84 continuing students from this number, she finally had 127 students in her sample. The data from the items on the questionnaire were scored, totaled, and ranked. Bamberg concluded that the total amount of composition instruction in high school had increased from 1975 to 1979, and that the number of students in expository classes had doubled. This study illustrates the use of one type of sampling that will be explained below.

Research design: sample and survey

Sample selection

The first task of a researcher is to determine the large *population*, or *"N,"* from which a *sample*, *"n,"* is to be drawn. The target population for the Texas study was all collegiate writing programs in the United States. Because the researchers claimed that no central source listed all these programs or their directors, they used as a population 550 directors gleaned from the list of Writing Program Administrators and "others known to the directors." This population, however, turned out to be not as representative of the target population as they had anticipated. When the researchers compared characteristics of their sample group with known data about colleges from the National Center for Educational Statistics, they found some major differences in features like size and type of institution. Of their institutions, 37% were privately supported in contrast to 53% in the national population; 11% of theirs were two-year colleges and 36% were universities, while in the national population, the figures were 38% and 5%, respectively. About 5% of their institutions had fewer than 1000 students, while in the national population the percentage was 39. To have started with a more representative population, they could have obtained from the National Center for Educational Statistics, lists of addresses of all colleges and universities, that could be queried about their writing programs and directors. Eblen's population was 471 full-time faculty. Bamberg's was all students in beginning and advanced composition at UCLA.

When the population has been determined, researchers then decide on the size and type of sample by considering the question of feasibility: the number of units from which they can carefully collect good data and which they can adequately analyze. The precision of the information they learn about for the population is related to sample size. The *precision* of a sample refers to the "tightness" or relative certainty of the *confidence limits*.

Confidence limits

The confidence limits are the range of scores or percentages within which a population percentage is likely to be found on variables that describe that population. Table 4-1 is the table used to determine the percentage of confidence limits or precision of the projected sample

Table 4-1 Sample Size and Confidence Interval Limits[a]

Sample size	Confidence interval limits for percentages
10	±31%
20	±22%
30	±18%
40	±16%
50	±14%
60	±13%
70	±12%
80	±11%
90	±10.3%
100	±9.8%
150	±8.0%
200	±6.9%
250	±6.2%
300	±5.6%
400	±4.9%
500	±4.4%
1000	±3.1%

[a]The 95% confidence intervals based on a population incidence of 50% and a large population relative to sample size.

Based on $\dfrac{1.96 \sqrt{p \cdot q}}{\sqrt{n}}$

where $p = 50\%$ and therefore $q = 50\%$; $q = 1 - p$, or

$$\frac{1.96 \sqrt{2500}}{\sqrt{n}}$$

size regardless of population size, if the population is large. If, for example, a researcher finds after collecting data that in a sample size of 30 students, 60% have problems with organization, the researcher can conclude that in the entire population, 60% plus or minus 18% also have problems with organization. Sample sizes do not have to be large in proportion to the whole population to have precision. With a population of 6000, for example, a sample size of 240—1/25 of the population size (and 1/25 of the work in describing and analyzing it)—yields information that is about 1/5 as precise as the information from the whole population. A sample of only 60 from the same population (¼ of the prior sample) would yield information half as precise as the sample of 240. Thus, reasonably precise descriptive research can be done with far less effort and cost than if the entire population had to be studied. In 19 of 20 samples drawn at random from the population, the confidence interval, defined by the plus-or-minus confidence limits around the sample means of the results, will encompass the population mean. Sampling therefore reduces to manageable size with reasonable results what otherwise might have been an almost impossible descriptive study.

If a researcher, for example, wants to know the percentage of students with run-on sentence problems in the freshman population of a large university, Table 4-1 shows the level of precision that can be reached by various sample sizes. The level is shown in terms of plus-or-minus percentages for a given sample size. This table illustrates the worst case, the one in which the average incidence of run-ons in the population is 50%.

If the Texas researchers, for example, had analyzed a sample of 100 questionnaires from the population of writing program administrators and discovered that 42% of the institutions were giving proficiency exams, the researchers could have added and subtracted 9.8% to and from 42% (Table 4-1), creating a confidence limit from 32.2% to 51.8%. They could then have inferred that the percentage of proficiency exams used in the whole population was somewhere between 32.2% and 51.8%—the boundaries of the confidence limits.

CORRECTIONS IN CONFIDENCE LIMITS

After researchers have used Table 4-1 to determine the confidence limits, they sometimes need to add a correction to these limits if:

1. the sample size starts to approach the population size, or
2. the variable under scrutiny is known to have a *much* smaller or larger occurrence than 50% in the whole population.

Table 4-2   Correction Factors for Confidence
Limits When Sample Size (n) Is an Important
Part of Population Size ($N, \geq 100$)[a]

Sample percentage of population	Correction factor
5%	.98
10	.95
15%	.92
20%	.89
25%	.87
30%	.84
35%	.81
40%	.78
45%	.74
50%	.71
55%	.67
60%	.63
65%	.59
70%	.55

[a]For n over 70% of N, take all of N.

These correction factors are based on $\sqrt{\dfrac{N-n}{N-1}}$, where N
= population size, and n = sample size.

To make the correction for the first situation, researchers use the figure
in Table 4-2, multiplying it with the result obtained from Table 4-1. To
make the correction for the second situation, researchers use Table 4-
3. For the Texas research, the use of the correction in Table 4-2 was
unnecessary, because the sample size did not approach the population
size. The correction (Table 4-3) for an incidence markedly different from
50% was also unnecessary.

Precision and accuracy

Samples differ in the precision of their results, a variability primarily
due to sample size—the larger the sample, the smaller the variability
or the confidence limits. Table 4-1 presents the confidence limits in
terms of plus-or-minus value encompassing the observed average of a
characteristic in the sample. In the table, for a sample size of 100, the
plus-or-minus confidence limit is 9.8%. *Precision*, the tightness of the
confidence limits, differs from the *accuracy* of the sample. Even though

Table 4-3 Correction Factors for Table 4-1 [a]

Percentage incidence	Correction factor
50%	None
40% or 60%	.98
35% or 65%	.95
30% or 70%	.92
25% or 75%	.87
20% or 80%	.80
15% or 85%	.71
10% or 90%	.60
5% or 95%	.44
2.5% or 97.5%	.31

[a] Values for rare and common percentage incidence of variables.

These correction factors are based on $\sqrt{\dfrac{(P(1-P)}{2500}}$, where P = the estimated percentage of incidence in the population.

samples of various sizes are drawn at random from a designated large population of people or compositions or sentences, only rarely are the sample means identical to the population means from which they are drawn even though they are *unbiased*; in other words, the *means* (averages of scores) of random samples are rarely identical to the means of the population but are "expected to be" in the long run.

Choosing the size of a sample therefore involves balancing two factors: the degree of imprecision acceptable for decisions and conclusions, and the cost of collecting data and analyzing them. If the confidence limit of plus and minus 9.8% (for a sample size of 100) is too large for comfort, then the size of the sample will have to be increased within acceptable financial and time limitations. If the sample is increased to 500, for example, the confidence limits narrow considerably to plus and minus 4.4%; a sample size of 1000 yields confidence limits of plus and minus 3.1% (all given very large population sizes). However, the costs of collecting data are increased by a factor of five and ten, respectively, with a possible further loss in the quality of data.

When deciding on sample size, researchers need to think ahead in order to select a sample large enough to get the desired results for the group and possibly even subgroups. The sample size from a freshman class, for example, must be sufficient to allow for such subgroups as men in the freshman class or even the double cross-classification of freshman men who are athletes. Confidence limits should therefore be calculated with both the sample and subsamples in mind. Often much larger initial samples must be taken to get reasonable confidence limits on subsamples.

Random sampling

The simplest and often best way to select as representative a sample as possible is to use *random sampling*. (The term *random* signifies having no regular discernible pattern or order. See also Appendix I.) Random sampling is a straighforward process. If the population is not already numbered with something like sequential student numbers, the first step is to number the population. For example, if one wants to randomize students in several sections of Freshman Composition, one could start with the first section at 8:00 A.M., number the students in alphabetical order from one to twenty-five, and then, in the second section, resume numbering with twenty-six, etc. The order is not important as long as all students are numbered. Once the population is numbered, the researcher consults a table of random numbers to draw the requisite sample (see Table 4-4).

Randomizing involves the following points:

1. In advance the researcher decides three things arbitrarily: the point at which to start in the table, and the direction to read (e.g., typically down the first set of columns and then down the next set of columns to the right).
2. The researcher reads a sufficient number of the digit numbers to cover the total population (e.g., two numbers across for a population numbering within 1–99, three numbers across for a population of 999, and so forth).
3. When more than a single column of random numbers is used (for sample sizes of ten or greater), the numbers in the new columns generally do not overlap with those in prior columns.
4. When a page of random numbers has been exhausted, then the next page is used, again typically starting at the upper left set of columns.
5. Duplicate random numbers are skipped as are all random numbers beyond the range of the sample's highest number.
6. When large samples are to be drawn, it is best to use a large set of random numbers so that pages are not used more than once.

Random numbers can also be generated by a calculator or computer with a random-number function. Manuals provide directions that yield a sequence of random numbers in the display panel. These may have

Table 4-4 Random Numbers for Use in Drawing Random Samples

22 17 68 65 84	68 95 23 92 35	87 02 22 57 51	61 09 43 95 06	58 24 82 03 47
19 36 27 59 46	13 79 93 37 55	39 77 32 77 09	85 52 05 30 62	47 83 51 62 74
16 77 23 02 77	09 61 87 25 21	28 06 24 25 93	16 71 13 59 78	23 05 47 47 25
78 43 76 71 61	20 44 90 32 64	97 67 63 99 61	46 38 03 93 22	69 81 21 99 21
03 28 28 26 08	73 37 32 04 05	69 30 16 09 05	88 69 58 28 99	35 07 44 75 47
93 22 53 64 39	07 10 63 76 35	87 03 04 79 88	08 13 13 85 51	55 34 57 72 69
78 76 58 54 74	92 38 70 96 92	52 06 79 79 45	82 63 18 27 44	69 66 92 19 09
23 68 35 26 00	99 53 93 61 28	52 70 05 48 34	56 65 05 61 86	90 92 10 70 80
15 39 25 70 99	93 86 52 77 65	15 33 59 05 28	22 87 26 07 47	86 96 98 29 06
58 71 96 30 24	18 46 23 34 27	85 13 99 24 44	49 18 09 79 49	74 16 32 23 02
57 35 27 33 72	24 53 63 94 09	41 10 76 47 91	44 04 95 49 66	39 60 04 59 81
48 50 86 54 48	22 06 34 72 52	82 21 15 65 20	33 29 94 71 11	15 91 29 12 03
61 96 48 95 03	07 16 39 33 66	98 56 10 56 79	77 21 30 27 12	90 49 22 23 62
36 93 89 41 26	29 70 83 63 51	99 74 20 52 36	87 09 41 15 09	98 60 16 03 03
18 87 00 42 31	57 90 12 02 07	23 47 37 17 31	54 08 01 88 63	39 41 88 92 10
88 56 53 27 59	33 35 72 67 47	77 34 55 45 70	08 18 27 38 90	16 95 86 70 75
09 72 95 84 29	49 41 31 06 70	42 38 06 45 18	64 84 73 31 65	52 53 37 97 15
12 96 88 17 31	65 19 69 02 83	60 75 86 90 68	24 64 19 35 51	56 61 87 39 12
85 94 57 24 16	92 09 84 38 76	22 00 27 69 85	29 81 94 78 70	21 94 47 90 12
38 64 43 59 98	98 77 87 68 07	91 51 67 62 44	40 98 05 93 78	23 32 65 41 18
53 44 09 42 72	00 41 86 79 79	68 47 22 00 20	35 55 31 51 51	00 83 63 22 55
40 76 66 26 84	57 99 99 90 37	36 63 32 08 58	37 40 13 68 97	87 64 81 07 83
02 17 79 18 05	12 59 52 57 02	22 07 90 47 03	28 14 11 30 79	20 69 22 40 98
95 17 82 06 53	31 51 10 96 46	92 06 88 07 77	56 11 50 81 69	40 23 72 51 39
35 76 22 42 92	96 11 83 44 80	34 68 35 48 77	33 42 40 90 60	73 96 53 97 86
26 29 13 56 41	85 47 04 66 08	34 72 57 59 13	82 43 80 46 15	38 26 61 70 04
77 80 20 75 82	72 82 32 99 90	63 95 73 76 63	89 73 44 99 05	48 67 26 43 18
46 40 66 44 52	91 36 74 43 53	30 82 13 54 00	78 45 63 98 35	55 03 36 67 68
37 56 08 18 09	77 53 84 46 47	31 91 18 95 58	24 16 74 11 53	44 10 13 85 57
61 65 61 68 66	37 27 47 39 19	84 83 70 07 48	53 21 40 06 71	95 06 79 88 54
93 43 69 64 07	34 18 04 52 35	56 27 09 24 86	61 85 53 83 45	19 90 70 99 00
21 96 60 12 99	11 20 99 45 18	48 13 93 55 34	18 37 79 49 90	65 97 38 20 46
95 20 47 97 97	27 37 83 28 71	00 06 41 41 74	45 89 09 39 84	51 67 11 52 49
97 86 21 78 73	10 65 81 92 59	58 76 17 14 97	04 76 62 16 17	17 95 70 45 80
69 92 06 34 13	59 71 74 17 32	27 55 10 24 19	23 71 82 13 74	63 52 52 01 41
04 31 17 21 56	33 73 99 19 87	26 72 39 27 67	53 77 57 68 93	60 61 97 22 61
61 06 98 03 91	87 14 77 43 96	43 00 65 98 50	45 60 33 01 07	98 99 46 50 47
85 93 85 86 88	72 87 08 62 40	16 06 10 89 20	23 21 34 74 97	76 38 03 29 63
21 74 32 47 45	73 96 07 94 52	09 65 90 77 47	25 76 16 19 33	53 05 70 53 30
15 69 53 82 80	79 96 23 53 10	65 39 07 16 29	45 33 02 43 70	02 87 40 41 45
02 89 08 04 49	20 21 14 68 86	87 63 93 95 17	11 29 01 95 80	35 14 97 35 33
87 18 15 89 79	85 43 01 72 73	08 61 74 51 69	89 74 39 82 15	94 51 33 41 67
98 83 71 94 22	59 97 50 99 52	08 52 85 08 40	87 80 61 65 31	91 51 80 32 44
10 08 58 21 66	72 68 49 29 31	89 85 84 46 06	59 73 19 85 23	65 09 29 75 63
47 90 56 10 08	88 02 84 27 83	42 29 72 23 19	66 56 45 65 79	20 71 53 20 25
22 85 61 68 90	49 64 92 85 44	16 40 12 89 88	50 14 49 81 06	01 82 77 45 12
67 80 43 79 33	12 83 11 41 16	25 58 19 68 70	77 02 54 00 52	53 43 37 15 26
27 62 50 96 72	79 44 61 40 15	14 53 40 65 39	27 31 58 50 28	11 39 03 34 25
33 78 80 87 15	38 30 06 38 21	14 47 47 07 26	54 96 87 53 32	40 36 40 96 76
13 13 92 66 99	47 24 49 57 74	32 25 43 62 17	10 97 11 69 84	99 63 22 32 98

Source: Table 4-4 is taken from Table 33 of Fisher & Yates: *Statistical Tables for Biological Agricultural and Medical Research* published by Longman Group UK Ltd. London (previously published by Oliver and Boyd Ltd., Edinburgh) and by permission of the authors and publishers.

to be adapted to the problem at hand; in other words, no numbers over 50 may be able to be used, or more than two-digit numbers may be needed. In this case, arbitrarily adding sequentially 100, 200, or 300 systematically to the displayed number may be necessary. The ultimate object is to assure oneself that all units in the population have equal likelihood of being represented in the sample.

Through random sampling Eblen could have reduced her work to 118 questionnaires, 25% of the total number. To obtain a random sample, her first step would have been to number each of the faculty members, using the three-digit numbers from 001 to 999, to be sure that the 118 faculty members desired for the sample could be drawn in the sampling process.

By randomizing, the Texas researchers could have reduced the problem of securing their sample of 127 directors by means of self-selection through the willingness of directors to respond to the questionnaire. The investigators later compensated rather nicely for this problem by treating the data as a series of separate samples developed from a stratified sample (see below) in which every stratum was treated as a population in itself. To avoid these difficulties, however, they could have drawn a random sample from the list of colleges and universities published by the National Center for Educational Statistics.

In the process of random sampling many numbers may go unused. If researchers selecting a sample of 200 use three-digit numbers from 001 to 999, they would not need the numbers 201–999, thus wasting about 80% of the random numbers. If their sample were to be 600, they would not use almost 40% of the random numbers—those from 601 to 999—because the sample would run only from 001 to 600. But there is a way of using otherwise wasted time in drawing these unused numbers. If a researcher has a population of 200, multiples of 200 can be subtracted from each random number read larger than 200 (from 201 to 999), thus creating a much larger set of usable random numbers. The three digits "521" can be used by subtracting 200 twice, leaving the number 121 by which the 121st person, paper, or sentence would be selected in the sample. If Eblen had wanted a sample of exactly 100, she could have used the digits 00 to represent the number 100, thus narrowing the set of columns of random numbers she needed from three digits to two digits, covering her sample set from 00 to 99.

Random sampling, done strictly as directed, yields samples that are representative of the population from which they are drawn—within certain confidence limits. Had Eblen used a 25% random sample (118 faculty members out of 471), she would have had confidence limits of slightly larger than plus-and-minus 8.8%. With a sample size of 25% of her original population, she would have had a correction factor of .87, because her sample size starts to approach the population size (see Table 4-2). By taking a sample, she could have reduced her effort, perhaps acquired data sufficiently certain to make the point, and given herself more time to conduct interviews to follow-up questionnaires that remained unanswered.

Data collection

When researchers have selected optimum sample sizes, they are ready to collect data. What kinds of observations, questions, and instruments can be used in sample surveys? The answer is almost any kind of data collection method of interest to the researcher: questionnaires, paper collection, interviews, test results, and so forth.

Survey and questionnaire construction

Researchers often use questionnaires to collect data from large groups. If one intends to use a questionnaire, it is wise, for the first several attempts, to use questions that have been shown to be useful and non-ambiguous in prior studies. In the development of new questions, one can consult the extensive literature on developing survey questionnaires. (See References at the end of the chapter for some well-known titles.) It is imperative to edit questions for directness, simplicity, and clarity; to submit them for review by others; to use small pilot samples of the population of interest for reviews of initial response; and finally to edit and revise the questions, again based on the result of these reviews and tryouts, for use in the actual study. Care must also be taken in devising the cover letter to accompany questionnaires. (See References.)

Two basic types of questions used in questionnaires are objectively scored questions and open-ended questions. Multiple-choice questions are succinct, parsimonious, easily aggregated for analysis, and standardized, allowing the researcher to compare responses with those of other groups. Open-ended questions yield longer, more variable responses and provide less predetermined, more basic types of responses, but they are difficult to analyze and are limited by the writing ability and time of the respondents. When using open-ended questionnaires, researchers must be sure to pretest them to determine the time they require and the questions that are ambiguous or confusing. Even format needs to be checked. In one recent study, a researcher used a two-page set of questions without pretesting it. About twenty-five respondents failed to answer the questions on the back of the first page.

Other problems afflict questionnaires; the questions can

1. be noisy (too wordy or complex) or ambiguous;
2. ask for information that respondents cannot answer; and
3. pose suspicious questions.

Researchers must also decide whether the questions are appropriate and valid for gaining the desired information. Bamberg, for example, tried to ascertain the instructional content of high-school composition courses by relying on the recollections of college freshman. She asked them questions about how many of their classes had teacher explanations, class exercises, discussions, or assignments. This method relies heavily on students' memories and ability to discern types of English instruction, a method that seems less valid than more direct observation of the content and emphasis of high-school English classes.

The Texas researchers decided to develop a complex survey instrument that would elicit extensive and in-depth information, not just multiple-choice responses. Two outside experts estimated that it could be answered in an hour or two. Of the directors who were queried, 63.5% responded to these questions, providing lengthy statements testifying to the diversity of their writing programs, the student population served, and the curricula—all data contributing to the depth and individualization of the survey report. However, of the respondents, 10% complained that it took a minimum of three hours to answer questions and that some questions were poorly worded, confusing, or difficult to understand, suggesting that the questionnaire had not been pretested with writing directors independent of the project. The researchers identified another factor to account for the lower-than-expected return rate: the questionnaire arrived late in the spring semester, a very hectic period for composition directors.

Instead of observing classes herself, Eblen chose to have the faculty report on their course and classroom practices. She raised the question, however, of the accuracy of their self-reports and pointed out that the questions on classroom practice focused on the first class rather than over a semester.

Each of these studies shows the advantages and disadvantages of different methods of data collection. A prime consideration in the choice of method is its capability of eliciting high response rate.

RESPONSE RATES

Because all interpretations of averages, confidence limits, and other matters depend on following the sampling plan, nonresponses to questionnaires create a problem. Some nonresponse is likely, however, due to erroneous or out-of-date directories and mailing lists and particularly due to the failure of students, teachers, and administrators to respond to questionnaires or other survey instruments. To accommodate this lack of response, the sample should be somewhat larger to allow confidence limits to be based on the actual number responding.

In addition to confidence limits and administrative and fiscal feasibility, researchers should consider selecting samples small enough to be able to enhance the quality and response rate of the sample. Every effort should be made to reach and obtain responses from those identified in the sampling process. Sometimes two or three follow-ups and phone calls are needed to obtain these answers. At other times a smaller sample using face-to-face interviews or phone calls instead of mailed questionnaires will produce better quality data and far higher response rates. The goal should be at least an 80–90% response rate. It is also wise to gather information about each person in the sample in order to be able to check possible deviations of nonrespondents from the known characteristics of the population, in order to suggest how these deviations might influence the overall results. The excuse that a 40–50% rate to a mailed questionnaire is typical of the response rates in the field is not acceptable, because the accuracy (lack of bias) of the results of a sampling procedure is based on the sample's being representative of the entire population—the result of a strictly random sample of the population. No matter how large the return, a haphazard sample may be unrepresentative of the population of interest.

In the case of the Texas study, their praiseworthy effort to get an extraordinary amount of detailed, in-depth information brought its own set of limitations, primarily in the form of nonresponses to the questionnaire. They had a rate of return of 23% of the originally identified sample and 49% of those agreeing to respond, a response rate far too low to be representative. To increase their responses they could have, as indicated above, selected a small but representative random sample from the national list, rather than getting a sample by self-selection. That would have enabled them to use phone interviews or even visits to campuses, making possible the collection of more individualized in-depth data.

Their saving methodology was the foresight to collect data on the questionnaire that corresponded to known data in the higher education population of schools. When they determined that a clearly biased sample was in hand as the result of comparing these two sets of data, they divided the sample into three separate stratified samples (see below) based on the type of institution. This division made it possible to generalize more readily to populations of writing programs within the three school-type strata rather than attempting to generalize to all post-secondary collegiate institutions.

Data analysis

After sample and method selection, researchers determine the number of variables, the K, to be described. The data for the K variables are placed in a rectangular matrix (much like a teacher's class record book), with the sample units one to "n" down the left side and the variables, one to K across the top. The rectangle of data cells thus formed is called an n by K data matrix (see Figure 4-1). Unlike case studies in which the n is small and the K large, in sample survey studies the n is much larger than the K, because the researcher wants to study a few features of a large group that has been reduced to a sample by sampling procedures.

NOMINAL, INTERVAL DATA, AND RANK ORDER DATA

Three types of data may be collected in a survey: *nominal* data, *interval* data, and *rank order* data. Nominal data are those that result from simple counting: the total number of comma faults in a composition or a somewhat more refined set of data—the percentage (resulting from two or more sets of counts) of embedded or unembedded sentences. Witte and fellow researchers collected frequency data for different types of institutions: the number of required and elective, introductory and nonintroductory writing courses; the number of sections and students in each section; the number of full-time, part-time, graduate student, and tenure-track faculty teaching these sections; the number and type of textbooks used; the number of proficiency and exit examinations; and the types of writing activities, purposes, and papers required in these courses. Each of these categories was nominal. Eblen also collected frequency data: the number of different writing qualities that instructors expected, the number of types of problems they perceived, and the number of kinds of writing they required. For these nominal variables, each study presented both numbers and percentages. Frequency data presented in percentage or proportion form are typically

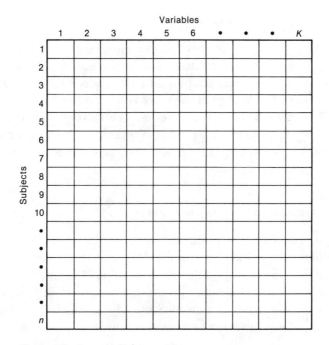

Figure 4-1 An $n \times K$ data matrix.

more convenient in sampling research studies. Because surveys are often conducted to find the percentage of some characteristic, frequency data, expressed as percentages, are doubly useful.

Interval data typically come from test scores with large numbers of items, from ratings, or from grades. Witte et al. also collected a variety of interval data including ratings of professors on different means of evaluating student writing and on successful or unsuccessful aspects of writing programs. Even though a researcher can calculate confidence limits or the precision of results for interval data, these results cannot be so readily tabled as is done when the confidence limits are expressed as plus-or-minus percentages (around an average percentage value). Thus, many sampling studies are designed to collect nominal data that can be expressed as percentages.

A special type of interval data is *rank order data* in which subjects, compositions, or teachers, are placed in a hierarchical order and assigned ranks from 1 to n. Because these rank orders are often assumed to be equal intervals, rank order statistical analyses (nonparametric) are used. The confidence limits of rank order data are not as readily tabled as percentages from nominal data.

MEANS

In addition to giving the counts and percentages of the variables being studied, a researcher is often interested in presenting sets of scores by calculating *means* (see next chapter and Appendix I for a fuller discussion). A *mean* is the average of the scores on whatever measure is being applied. For example, Bamberg gave the mean scores for English 1 and Subject A on the composition index; Witte et al. provided the mean number of pages written in first- and second-semester courses in different institutions.

RANGE, STANDARD DEVIATION, AND VARIANCE

The range, standard deviation, and variance are all measures of individual differences or indicators of the degree of variability of people on a variable. The *range* is the highest score minus the lowest score on a variable in a sample. A *standard deviation* is about one-fifth to one-sixth of the range. The *variance* is a measure of the amount of individual differences among the scores and is the square of the standard deviation.

Precision of data, a major concern discussed above, depends on the amount of variability of each characteristic in the population studied. For nominal data expressed in terms of percentages, the standard deviation is derived from the mean. For interval data, the precision is related to the standard deviations on the variables in the population. Many studies go beyond calculating means and standard deviations to determine whether the relationships among variables are "significant," that is, ones that probably would not be due to chance alone (see Appendix I).

Other types of sampling

Systematic random sampling

The theory and method of random sampling can also work well for other types of samples like systematic sampling, which is useful when the population to be studied (students or sentences in a set of compositions) is already organized in a sequence in the data. The important concern is to establish an order in advance to account for all units to be studied. Sentences in a discourse, for example, are already in order. Once the sequential order is established, every "Nth" sampling unit (say fifth, or tenth, or twentieth) is observed. The interval is set on the basis of sample size desired; in other words, larger intervals produce smaller sample sizes. But more important concerns exist. First, the observations should start at a point picked at random within the sequential unit. For example, a researcher deciding to choose every tenth sentence in a set of papers would select a number from 1 to 10 and make the first observation on whatever number was randomly chosen (e.g., 3). In this case, the researcher would look at sentences 3, 13, 23, etc. Second, the researcher must examine the population for any natural or planned order in it. Have the students alternated simple with complex sentences? If so, then if the researcher happens to choose an even number as the interval, only simple sentences would be examined. Generally, the interval unit should not be related to any ordinary sequence in the data. Third, the entire population must be covered; in other words, the size of the population, when divided by the size of the sampling unit, should about equal the desired sample size. If systematic random sampling is possible, the task of selection can be done more expeditiously than in regular random sampling.

One way to sample people systematically is to take all those with social security numbers ending with a digit drawn at random, say 4. This yields a sample one-tenth the size of the population. A one-fifth sample size can be developed by using two different last digits of the social security numbers. A 5% sample can be taken by drawing at random five two-digit random numbers and comparing them to the last two digits. (Provisions have to be made, however, for foreign students or those who have only worked in civil service, etc.) All of these systematic samples are treated analytically as if they were random samples.

Quota sampling

Quota samples are helpful when a researcher knows the percentage of incidence of specific features of the population like sex of students, race, or college level. In the Texas study, the researchers, consulting the National Center for Educational Statistics, found that 53% of the schools in the national population were private, 38% were two-year colleges, and 5% were universities. In order to obtain a reasonably representative sample from the whole population, they would have had to select programs that filled those same percentages. In quota sampling, therefore, a population and some known characteristics are identified and then samples are picked so that the ultimate sample matches those known characteristics (e.g., percentage of male and female, percentage of various minorities, and percentage of upperclassmen).

Stratified samples

Stratified samples are used when some parts of the population are of more immediate interest than others. For instance, if the instructional content of freshman composition were being developed, the researcher might want to examine freshman students in more detail than upperclassmen. The college population would therefore be divided into two strata: freshman and others. Essentially each of these strata becomes a population in itself, and random, systematic, or quota samples can be taken from each strata. Eblen could have used stratified sampling, selecting a designated number of faculty members from each of the five academic units.

Stratified sampling allows one to observe a larger sample of the kind of unit under scrutiny, and possibly a larger set of characteristics, than if one sampled the entire student body or entire faculty. If, after studying each stratum separately, the investigator wants to examine characteristics of all the strata, several strata can then be combined (with appropriate weighting) to get a picture of the whole population. Care should be taken not to use strata that will end up with small numbers as the result of numerous cross-classifications. For example, the stratum of freshmen on varsity athletic teams may yield a population that is miniscule and a sample that is almost nonexistent. The confidence intervals as the result of the very small sample would be very large—useless for most purposes.

The Texas study, as we indicated, compensated for its low response rate by treating separate strata (e.g., four- and two-year institutions) *ex post facto* as populations in themselves from which they extracted samples. They made weighted extrapolations to a more "normalized" U.S. population by number of institutions and by enrollment, but they suggested that the projections should be viewed with caution.

Sometimes stratified samples have large confidence limits if the sample within the strata is too low. What confidence limits did the Texas study have around the responses within the three strata? In their table (p. 88), "Percentages of Institutions Offering Workshops for Faculty Groups for Improving Composition Teaching," twelve schools from the two-year college stratum responded, of which 58.3%, or seven schools, offered workshops for part-time faculty. In the table of confidence limits (see Table 4-1), this sample size of twelve is somewhat larger than the tabled value of ten, which has confidence limits of plus-and-minus 31%. If this percentage is added to the sample mean of 58.3%, the confidence limits are 27.3% and 89.3%. (This is slightly larger than the actual precision of this part of the Texas study.) Thus, even if the sample were representative of the population, one would expect the true population percentage of two-year schools offering workshops to be somewhere in the range of the confidence limits between 27.3% and 89.3% in 95 out of 100 random samples.

In their biggest sample from the stratum of four-year colleges, 63.5% of these sample institutions gave workshops for tenured full-time faculty. With a sample of 40 institutions, the plus-and-minus confidence interval values from Table 4-1 were 16%, creating a lower confidence limit value of 47.5% and a higher limit of 79.5%. Perhaps the cross-classification of the data was overdone, given the sample sizes.

Cluster samples

Another method, cluster sampling, has been used by researchers who wish to study individual units (students or compositions) within a large population such as a school, but who do not want to draw them randomly because of the difficulty in studying students scattered in different classes. In these cases, researchers have drawn samples of whole classes of students on which to make observations. Because it may be difficult to assure the representativeness of such samples and especially to determine the confidence limits, cluster sampling should be avoided unless one has a statistician as a design consultant. Even then it may be difficult to obtain good estimates of the confidence limits because the relationships between the clustering variables and other variables are not well known. (Note, however, that cluster sampling is used by many good national sampling organizations, which, after consulting with statisticians to estimate the costs and benefits of various sampling methods, have chosen cluster sampling.)

Bamberg's use of cluster sampling illustrates some of the strengths and weaknesses of this kind of sampling. Her first difficulty, which she carefully explained in her report, came because of a shifting population. The 1975 UCLA composition program consisted of three courses: Subject B, a noncredit remedial course for students with very poor writing skills, Subject A, a noncredit course for students whose skills were slightly below an expected level for freshman, and English 1, the regular credit freshman English composition course. Bamberg did not include students in Subject B in her 1975 study. In 1978, UCLA shifted to only one noncredit course, English A, followed by English 1 in place of the former Subject A, and English 3 in place of the former English 1. Because criteria for assigning students to these courses also changed, the populations for the two studies were not the same. The new English 1 included some students who might have been in Subject A and excluded the better writers who were exempt. Bamberg selected her sample from these populations on the basis of samples of *sections* at class hours but then analyzed *students* in these classes. Her analysis units, therefore, were students, but her samples were classes, making her confidence limits difficult to assess. The confidence interval, based on number of students rather than number of classes, was rather good—about plus-and-minus 8.8%—for the sample of 127. But note that this was probably an underestimate because the number of classes was smaller than the number of students. The use of a cluster sample, especially with possible homogeneous groups of students in the classes, made it difficult to estimate the sampling limits.

Despite these problems, she tried to bring a major task, that of describing high-school English teaching methods in California, within the realm of possibility by using a sampling method. She related a set of characteristics of a large population, California high schools, to a relatively small number of major variables within certain limits that were inevitably less than precise because of cost limitations. She described in reasonable detail the number of courses and the type and emphasis of high-school course work in English composition of beginning and advanced freshman composition students at UCLA in 1975 and again in 1979. She compared these features to the major variables—English 1 and 3, in 1975 and 1979.

Results

Sample surveys are descriptive research from which it is difficult to make cause-and-effect statements. A researcher must be careful, therefore, about drawing such conclusions about results. Another caution pertains to representativeness. If random sampling has been used, a researcher can extend conclusions, within the appropriate confidence limits, to the large population from which the sample was drawn. Cluster sampling, because of the problems with confidence limits, complicates a researcher's task of extending conclusions. To the extent that nonrandom techniques are used, samples are likely to be atypical of the population unless effective compensation occurs.

Was Bamberg's sample typical of the total amount and type of com-

Table 4-5   High School Composition Instruction of Freshmen Composition Students

Aspects of writing	Less Than Three Semesters of Instruction		More Than Three Semesters of Instruction		χ^2	p
	1975 ($N=288$)	1979 ($N=127$)	1975 ($N=288$)	1979 ($N=127$)		
Content Development and Organization						
(a) Stating thesis	47%	32%	53%	68%	8.92	.002
(b) Supporting ideas	51%	35%	49%	65%	8.26	.0004
(c) Organizing essay	48%	28%	52%	72%	15.50	.0001
(d) Paragraphing	61%	56%	39%	44%	.86	.35
Correct Written Form						
(e) Correct grammar and punctuation	44%	33%	56%	67%	4.71	.02
(f) Spelling	54%	38%	46%	62%	9.42	.002
Writing Style						
(g) Choosing the right word	66%	65%	46%	35%	.07	.78
(h) Revising sentence structure	58%	63%	42%	37%	.79	.37
(i) Being concise	60%	65%	40%	35%	1.18	.27

Source: Bamberg (1981), Table 1. Copyright © 1981 by the National Council of Teachers of English. Reprinted by permission of the publisher.

position instruction in the California high schools? The sample, UCLA freshmen "who represented the top 12.5 percent of California high school graduates" (1981, p. 259), cannot be said to be representative of California high-school students in general. If she wanted to generalize to all California high schools, should she not have asked questions at several other colleges?

Her data yielded some interesting conclusions (see Tables 4-5 and 4-6). The mean percentages on "Aspects of Writing" of students with less than three semesters and those with more than three semesters were *not* significantly different on Paragraphing, Choosing the Right Word, Revising Sentence Structure, or Being Concise. These means suggest several possibilities: these high school practices actually hampered students' later writing, the questionnaire's reliance on memory was not highly valid, or students with more than three semesters of instruction spent more time on higher level composition tasks. The last was probably most likely.

Because Eblen did not select a sample by any of the means discussed above, her sample came from the self-selection of those who responded. The sample sizes reported for the five divisions were 54, 26,

Table 4-6 Mean Scores on Composition Index and Components for 1975 and 1979 Freshmen Composition Students

	Mean			
	1975 (N = 288)	1979 (N = 127)	t	p
Composition Instruction Index	16.50	18.20	2.62	.009
Index Components				
Content Development and Organization	6.36	7.26	3.04	.002
Correct Written Form	3.18	3.81	3.68	.000
Writing Style	4.12	3.81	−1.39	.16
Expository Writing Practice	2.82	3.34	3.31	.001

Source: Bamberg (1981), Table 2. Copyright © 1981 by the National Council of Teachers of English. Reprinted by permission of the publisher.

53, 61, and 68. The confidence limits, therefore, on the various percentages reported were at best about 12% for the largest sample and on the order of plus-and-minus 20% for the smallest sample, certainly not highly precise and indicative of an overanalysis and reporting of data. Eblen also seems to have extended her conclusions to "the whole faculty," a generalization that is risky to make because she did not use a sampling procedure and received only a 56.6% return rate. She could only form conclusions about those faculty who responded to the questionnaire. She rightly suggested that faculty who responded may have been more apt to work with student writing than those who failed to respond.

National surveys

Surveys of national, regional, or statewide populations require organizations capable of assessing samples large enough to obtain reasonably precise confidence intervals and an 80–90% response rate. In fact, they usually necessitate the sponsorship of a national organization or funding agency to be done well. When too many individuals attempt such surveys, they discourage responses to major efforts. In the Texas report, one can see the frustration of a writing program director who said: "I am ordinarily a cooperative and dogged filler-out of forms and questionnaires. This past year alone I count at least seven full-length efforts, not all of which I considered very useful" (p. 112). He had a good point. Clearly in that year there could not have been seven major national composition surveys.

National and regional sampling survey studies should therefore seldom be done, and then only with substantial support. Only those well trained in questionnaire design and sampling procedures should use them for national studies. But this does not mean that researchers cannot work with national data. They can add questions on composition to general national surveys that are continually being taken, gaining the advantage of interrelationship with other questions. The National Center for Educational Statistics, for example, has as its major function the development and analysis of data generated from national questionnaires. It conducts a continuous study of students in their sophomore and senior years and then follows them for several years beyond graduation. These studies have excellent response rates, even after high school, because they often use Census Bureau in-person interviews to obtain the data. To these national instruments, composition researchers can add questions that have been carefully screened by review panels and can obtain tapes of the complete data set for intensive analysis. They can also do more detailed analyses and reporting of the responses to the questions asked in their particular areas of interest.

Since 1969, the National Assessment of Educational Progress has surveyed the progress of students ages nine, thirteen, and seventeen in ten different learning areas including writing. (This project is now conducted by the Educational Testing Service.) Even though questions cannot be added to these studies, the researcher has access to a wealth of longitudinal data about writing in relation to many other variables. In 1975, Mellon, for example, used the National Assessment writing data to compare the 1968 and 1974 samples, concluding that there was no evidence of decreased writing ability in 1974. Other data about writing are also available. Project TALENT (1972) collected writing samples from thousands of adolescents in conjunction with some 500 other variables. While extensive analyses and reports on these survey data have been made, the bulk of the information available from the writing samples is still untouched.

A new, large data base is now being created by the International Association for the Evaluation of Educational Achievement, which is studying students' writing in seventeen countries. An international group of researchers is collecting student papers on nine writing tasks taken from three age levels—end of primary education, end of compulsory education, and end of secondary education. These papers encompass a range of purposes and audiences and cognitive tasks (Purves and Takala, 1982).

Summary

In general, sampling and survey studies provide a valuable means of obtaining representative descriptive information about the writing of very large populations because sampling procedures reduce them to manageable size. But the precision of the information is directly related to sample size. Although true representativeness requires that a sample be chosen randomly, several approximations are available such as systematic, quota, and stratified sampling. Populations, samples, and sampling units must be well-defined. The instruments used to collect the data must be chosen with considerable care. If available, the researcher should use prior questionnaire items and observational methods known to be valid and appropriate for the intended population. If new instruments need to be developed, they must be reviewed, edited, and tried on representative populations before being used. Finally, generalizations must be made only to the population from which the sample was drawn. This descriptive research design, if conducted carefully, can transform unwieldy projects into manageable research and yield valuable, representative descriptive data for composition studies.

Checklist for reading "sampling survey studies"

1. What was the population, the N? How and why was it selected?
2. What was the sample? What method of sampling was used? Why?
3. What were the confidence limits? Are they acceptable for the use to which the results will be put?
4. On what theory were the questionnaire or survey items based?
5. Were the questions clear and unambiguous?
6. What kind of data (nominal, interval, or rank order) were collected? How?
7. What were the K variables?
8. What percentage of returns was obtained?
9. What conclusions were drawn?

References

Research Design and References

Anderson, P.V. (1986). Survey methodology. In L. Odell & D. Goswami (Eds.). *Writing in nonacademic settings.* New York: Guilford Press, pp. 453–498.

Asher, J.W. (1976). *Educational research and evaluation methods.* Boston: Little, Brown.

Belson, W.A. (1981). *The design and understanding of survey questions.* Aldershot Hants, England: Gower.

Berdie, D.R., & Anderson, J.F. (1974). *Questionnaires: Designs and use.* Metuchen, N.J.: Scarecrow Press.

Bradburn, N.M. (1979). *Improving interview method and questionnaire design.* San Francisco: Jossey-Bass.

Burkland, J., & Grimm, N. (1984). *Students' responses to our response: Parts I and II.* ERIC Document Reproduction Service No. ED 254 241.

Converse, J.M., & Schuman, H. (1974). *Conversations at random: Survey research as interviewers see it.* New York: Wiley.

Dillman, D.A. (1978). *Mail and telephone surveys: The total design method.* New York: Wiley.

Fink, A., & Kosecoff, J. (1985). *How to conduct surveys.* Beverly Hills, Calif.: Sage Publications.

Fisher, R.A. & Yates, F. (1974). *Statistical tables for biological, agricultural and medical research.* 6th ed. London: Oliver and Boyd.

Flanagan, J.C. (1972). Project TALENT data bank: A handbook. American Institute for Research, P.O. Box 1113, Palo Alto, Calif.

Fowler, F.J. (1984). *Survey research methods.* Beverly Hills, Calif.: Sage Publications.

Frey, J.H. (1983). *Survey research by telephone.* Beverly Hills, Calif.: Sage Publications.

Garrett, A. (1972). *Interviewing: Its principles and methods.* New York: Family Service Association of America.

Gorden, R.L. (1975). *Interviewing: Strategy, techniques, and tactics.* Homewood, Ill.: Dorsey Press.

Guilford, J.P., & Fruchter, B. (1973). *Fundamental statistics in psychology and education,* 5th edition. New York: McGraw-Hill.

Hansen, M.H., Hurwitz, W.N., & Madow, W.G. (1953). *Sample survey methods and theory* (Vol. 1). New York: Wiley.

Kish, L. (1965). *Survey sampling.* New York: Wiley.

Labaw, P.J. (1980). *Advanced questionnaire design.* Cambridge, Mass.: ABT Books.

Mellon, J. (1975). *National assessment and the teaching of English.* Urbana, Ill.: National Council of Teachers of English.

Oppenheim, A.N. (1966). *Questionnaire design and attitude measurement.* New York: Basic Books.

Parten, M.B. (1950). *Surveys, polls, and samples.* New York: Harper & Row.

Schuman, H. (1981). *Questions and answers in attitude surveys: Experiments on question form, wording, and content.* New York: Academic Press.

Seidman, S., & Bradburn, N.M. (1982). *Asking questions.* San Francisco: Jossey-Bass.

Sonquist, J.A. (1977). *Survey and opinion research: Procedures for processing and analysis.* Englewood Cliffs, N.J.: Prentice-Hall.

Stephen, F.J., & McCarthy, P.J. (1963). *Sampling opinions.* New York: Wiley.

Williams, W. (1978). *A sample of sampling.* New York: Wiley.

Witte, S., et al. (1981). *The empirical development of an instrument for reporting course and teacher effectiveness in college writing classes.* Technical Report No. 3. Fund for the Improvement of Postsecondary Education. University of Texas, Austin.

Survey and Sampling Studies

Andersen, P.V. (1986). What survey research tells us about writing at work. In L. Odell & D. Goswami (Eds.). *Writing in nonacademic settings.* (pp. 1–83). New York: Guilford Press.

Applebee, A. (1982). Teaching high-achieving students: A survey of the winners of the 1977 NCTE Achievement Awards in Writing. *RTE, 12,* 339–348.

Bamberg, B. (1978). Composition instruction does make a difference: A comparison of the high school preparation of college freshmen in regular and remedial English classes. *RTE, 12,* 47–59.

Bamberg, B. (1981). Composition in the secondary English curriculum: Some current trends and directions for the eighties. *RTE, 15,* 257–266.

Boiarsky, D. (1984). What authorities tell us about teaching writing: Results of a survey of authorities on teaching writing. Eric Document Reproduction Service No. ED 243 145.

Bossone, R.M, & Larson, R.L. (1980). *Needed research in the teaching of English.* New York: Center for the Advancement of Study in Education.

Bridgeman, B., & Carlson, S. (1984). Survey of academic writing tasks. *Written Communication, 1,* 247–280.

Brinkley, W.J. (1984). A comparison of supervisory-level managers' written business communication practices and problems and college written business communication instruction. *DAI, 45,* 08A.

Cotton, H.S. (1984). The teaching of creative writing in selected colleges and universities, 1970–1980: Issues, activities, and trends. *DAI, 45,* 12A.

Daly, J. (979). Writing apprehension in the classroom: Teacher role expectancies of the apprehensive writer. *RTE, 13,* 37–44.

Eblen, C. (1983). Writing across the curriculum: A survey of a university faculty's views and classroom practices. *RTE, 17,* 343–348.

Ede, L., & Lunsford, A. (1986). Collaborative learning: Lessons from the world of work. *Writing Program Administrator 5,* 71–77.

Fagan, E.R., & Laine, C.H. (1980). Two perspectives of undergraduate English teacher preparation. *RTE, 14,* 67–72.

Faigley, L. & Miller, T. (1982). What have we learned from writing on the job? *College English, 44,* 557–569.

Gillis, C. (1977). A report on the *EJ* readership survey. *English Journal, 66,* 20–26.

Halpern, J. (1986). An electronic odyssey. In L. Odell & D. Goswami (Eds.). *Writing in nonacademic settings.* (pp. 155–202). New York: Guilford Press.

Hanon, K.W. (1984). Text and context: A study of business writing strategies. *DAI, 45,* 04A.

Herrington, A.J. (1985). Writing in academic settings: A study of contexts for writing in two college chemical engineering courses. *RTE, 19,* 331–361.

Hillocks, G. (1972). *Alternatives in English: A critical appraisal of elective programs.* Urbana, Ill.: National Council of Teachers of English.

Hoetker, J., & Brossell, G. (1980). *EJ* readership survey: A report. *English Journal, 69,* 13–20.

Hogan, T. (1980). Students' interest in writing activities. *RTE, 14,* 119–125.

Jablin, F.M., & Krone, K. (1984). Characteristics of rejection letters and their effects on job applications. *Written Communication, 1*(4), 387–406.

Jeffrey, C. (1981). Teachers' and students' perceptions of the writing process. *RTE, 15,* 215–228.

Jordon, K.N. (1984). A study of language attitudes of selected teachers K–12 in five Arkansas communities. *DAI, 45,* 02A.

King, B.L. (1979). Two modes of analyzing teacher and student attitudes toward writing: The Emig Attitude Scale and King Construct Scale. *DAI, 40,* 07A.

Lunsford, A. (1977). An historical, descriptive, and evaluative study of remedial English in American colleges and universities. *DAI, 38,* 2743A.

Lunsford, A., & Ede, L. (1986). Why write . . . together: A research update. *Rhetoric Review, 9,* 10–26.

Lynch, C., & Klemans, P. (1978). Evaluating our evaluations. *College English, 40,* 166–180.

Meyers, C.D. (1984). Exploring differences in cognitive strategies among and between Freshman English students and instructors: A research model. *The Writing Instructor, 3,* 61–74.

Moore, W.J. (1984). Criteria and consistency of freshman composition evaluation: A national survey. *DAI, 45,* 07A.

Noreen, R. (1977). Placement procedures for freshman composition: A survey. *CCC, 28,* 141–144.

O'Donnell, R.C. (1979). Research in the teaching of English: Some observations and questions. *English Education, 10*(3), 181–183.

Pearce, D.L. (1984). Writing in content area classrooms. *Reading World, 23*(3), 234–241.

Pollay, R. (1984). Twentieth-century magazine advertising. *Written Communication, 1,* 56–77.

Purves, A., & Takala, S. (Eds.) (1982). *An international perspective on the evaluation of written composition.* London: Pergamon.

Shatshat, H. (1980). A comparative study of the present and ideal roles of communication directors in selected business organizations. *Journal of Business Communication, 17*(3), 51–63.

Squire, J., & Applebee, A. (1966). *A study of English programs in selected high schools which consistently educate outstanding students in English.* Project No. 1994. University of Illinois, Urbana.

Tighe, M.A., & Koziol, S. (1982). Practices in the teaching of writing by teachers of English, social studies, and science. *English Education, 14,* 76–85.

Westerfield, M.W. (1984). The behaviors and attitudes of non-English faculty at York College toward the teaching of writing. *DAI, 45,* 07A.

Williams, J.D., & Alden, S.D. (1983). Motivation in the composition classroom. *RTE, 17,* 101–112.

Witte, S., Meyer, P., Miller, T., & Faigley, L. (1981). *A national survey of college and university writing program directors.* Technical Report No. 2. Fund for the Improvement of Postsecondary Education Grant No. G008005896. University of Texas, Austin.

Witte, S., Cherry, R., & Meyer,, P. (1982). *The goals of freshman writing programs as perceived by a national sample of college and university writing program directors and teachers.* Technical Report No. 5. Fund for the Improvement of Postsecondary Education. University of Texas, Austin.

Witte, S., Meyer, P., & Miller, T. (1982). *A national survey of college and university teachers of writing.* Technical Fund for the Improvement of Postsecondary Education. University of Texas, Austin.

Zemelman, S. (1977). How colleges encourage students' writing. *RTE, 11,* 227–234.

5

Quantitative
Descriptive Studies

As we explained in Chapters 2 and 3, case studies and ethnographies describe a few people and situations on all variables of theoretical importance and interest to the researcher. The purpose of these qualitative studies is to generate variables, to operationally define them, and to develop early understandings of their interrelationships. These findings all merge into tentative theories or sets of hypotheses that form the first phase of all research in the effort to *understand* and to *explain* any phenomena.

Quantitative descriptive research goes beyond case studies and ethnographies to isolate systematically the most important variables developed by these studies, to define them further, to *quantify* them at least roughly, if not with some accuracy, and to interrelate them. Researchers observe certain defined and quantified variables for a reasonably large number of subjects over either a short period of time or a series of time intervals. They then correlate these variables by various statistical means in order to see whether strong, weak, or no relationships emerge. In our taxonomy of designs, quantitative descriptive research goes beyond thick description and variable identification; it reports the results of statistical analyses on these variables. But it is descriptive, not experimental research, because no control groups are created and no treatments are given.

The following studies will illustrate the principles, procedures, and some of the statistical analyses common to this kind of research.

Three quantitative descriptive studies

Study 1 Sharon Pianko (1981). A description of the composing processes of college freshmen. *RTE, 13,* 5–22.

Pianko did a quantitative descriptive study of twenty-four community college students (half traditional and half remedial, half adult and half typical college entrance age) to see "if different groups of college writers follow the same patterns as those of younger writers and to see if there are other ways of characterizing the writing processes for different types of students" (pp. 5–6).

Study 2 Lee Odell and Dixie Goswami (1982). Writing in a non-academic setting. *RTE, 16,* 201–223.

Odell and Goswami studied the reasons for writing choices made by nonacademic writers. They were interested in the extent to which nonacademic writers were sensitive to rhetorical issues such as the relationship to audience and to the ethos conveyed in their writing. They also wanted to see whether stylistic choices varied according to audience, purpose, and types of discourse and whether writing varied from one group to another within a given institution. To investigate these questions, they studied five administrators and six caseworkers in a county social service agency.

Study 3 Paul Diederich (1974). *Measuring growth in English.* Champaign, Ill.: National Council of Teachers of English.

Diederich and his associates at Educational Testing Service wanted to improve the reliability of grades on essays. To examine the factors operative in judging writing ability, they obtained 300 papers written by students in their first month at three different colleges. These papers were graded by sixty distinguished readers: college English teachers, social science teachers, natural science teachers, writers, editors, lawyers, and business executives. Because seven were unable to complete the grading, the study was conducted with fifty-three readers. The study isolated five factors used by readers to judge the essays: ideas, organization, wording, flavor, and conventions (usage, punctuation, spelling, and handwriting).

Research design: quantitative descriptive studies

Subject selection

Subject selection in quantitative descriptive research depends on the number and type of variables the researcher wants to study. Because the variables will be quantified, researchers need a larger number of subjects than either case study or ethnography uses. Generally a safe number is at least ten times as many subjects as variables. Using a small number of subjects or using a ratio of subjects to variables of less than ten to one may cause the researcher to overinterpret possible relationships—in other words, to capitalize on relationships that could occur by chance alone. When samples are small and measurement instruments brief, however, the researcher will find it difficult to obtain relationships that meet traditional levels of statistical significance (i.e., 5% and 1%).

Subjects should be selected for their appropriateness to the variables under scrutiny and their availability. Pianko had access to a community college freshman population that was in part older than many freshmen and had a high percentage of students in remedial classes. Her selection was guided by her theoretical need to study the composing processes of college students of various backgrounds and ages. She nicely blended her theoretical questions with the available population. Odell and Goswami selected their sample of eleven subjects from a nonacademic environment to which they had access. They chose their five administrators and six caseworkers on the basis of availability and interviews with thirty workers identified by informants. Diederich chose 300 papers from three different colleges (unidentified) and sixty readers from different fields in order to isolate the factors that educated readers use to grade essays.

If researchers want to add representativeness to the study, they can use any of the sampling techniques discussed in Chapter 4. Pianko selected at random 24 students from 400 enrolled in a community college freshman composition course. Her sample was a *quota sample* based on three dichotomous nominal variables: class status (traditional versus remedial), age (under 21 versus 21 and over), and sex. Because she used the sampling method, she could say that her results were representative of a much larger population of community college students. Unfortunately, as is the case in much research done in the real world of colleges, only 17 of Pianko's "24 randomly selected volunteers" remained through to the completion of the study.

Her ratio of students to variables was also a problem. The ratio rule would have required 220 subjects (based on 22 variables times 10), a massive task in the light of her descriptive detail and the effort needed to persuade 220 freshmen, instead of 17, to write five compositions and to submit to interviews and videotapes. As she concluded, the importance of her work clearly rested on the suggestions it gave about important relationships that need to be further studied with larger numbers of subjects.

Odell and Goswami used a sample of eleven subjects, which allowed them to do detailed analysis, but to gain more definitive insight into the meanings of their correlations, they needed more subjects—at least 10 times the variables. In Diederich's research, the use of 53 judges and 300 essays gave him the needed power to make meaningful correlations. Because the judges were from a wide variety of academic and nonacademic backgrounds (although not randomly selected from a population of well-defined readers), the study gained some measure of generality, which was enhanced by later replications of the study that yielded similar results.

Data collection

Data collection procedures can be multiple: any methods that give researchers the data that they need to quantify and interrelate the variables they wish to study. Pianko scheduled five writing episodes for each subject, one each week, in which subjects were asked to write a 400-word essay. They could take as much time as they wished on the assignment as long as they finished each one in an afternoon. Although the assignments were designed to elicit writing in each of four modes—description, narration, exposition, and argumentation—the majority of the writing (55%) was narrative. She observed and videotaped her students at least once during their writing, recording the length of time and frequency of certain behaviors. Following the completion of one of the assignments, she questioned students to obtain their views on the causes and meanings of certain composing behaviors. At the end of the entire series of writing sessions, she used a writing behavior question guide and a background interview guide to interview students about general attitudes, feelings, self-described behaviors, and past writing experiences, taping and transcribing all interviews. She also collected all pieces of writing: scraps, outlines, drafts, and the essays themselves.

Pianko's use of these methods (along with associated pieces of writing) gave the study a base of 22 quantitative descriptive variables. Her use of interviews provided qualitative data that yielded an in-depth understanding of the writing and thinking processes of this group. Her data collection was guided by a theoretical need for a comprehensive study of the composing processes of college students of various backgrounds and ages.

Odell and Goswami conducted initial interviews with their 11 subjects to determine the amount, types, and importance of writing in their jobs. They then conducted two subsequent interviews (two months apart) to elicit information about the bases for specific choices in their writing. They also collected writing from a two-week period, did a content analysis on it, and prepared a sheet for a third interview to determine more accurately the reasons for writers' choices. To obtain judgments about acceptability of style, they selected nine pieces of writing (three pink memos, three white memos, and three letters), prepared three versions of each piece, and had participants rate them as most, less, and least acceptable. To categorize the reasons given by writers for their choices, they had two judges read each transcript and categorize the reasons according to the variables they had identified. Three judges analyzed the linguistic features of the writing samples.

From the 53 judges, Diederich collected general merit ratings on a scale from 1 to 9. He also gathered from the readers brief comments about anything they liked or disliked in the papers. He obtained grades on 300 papers and 11,018 comments that were tabulated and grouped under 55 headings.

Variable selection

The most important aspect of quantitative descriptive study is the choice of variables to observe and intercorrelate. The source of variables is the theory of writing that researchers hold and wish to test. Sometimes the variables have been already identified in case studies and ethnographies, a fact that the literature search will disclose. In many studies, the variables are clustered into *independent* variables and *dependent* variables. We identify the terms *independent* and *dependent* here because they are used by many composition researchers. The distinciton between them, however, is rather imprecise in descriptive research. Researchers often call those variables independent which constitute differences in subjects prior to research—e.g., class level, age, or gender. Dependent variables are often those introduced by the researcher for the analysis, e.g., prewriting time, planning behavior, or number of pauses.

In Pianko's study the *independent variables* were

1. Instructional class type: remedial versus traditional
2. Age: under 21 versus over 21
3. Gender

The *dependent variables* were

4. Prewriting time
5. Composing time
6. Rate of composing (words per minute)
7. Rereading time
8. Revising (number of revisions per 100 words)
9. Number of pauses
10. Number of drafts per writing episode
11. Number of rescannings
12. Planning behavior
13. Attitude toward writing—positive/negative
14. Stylistic concerns
15. Considerations of purpose
16. Knowledge of ideas
17. Writer's concerns
18. Amount and type of self-initiated writing
19. Amount and kind of writing done in elementary and secondary schools
20. Amount and type of writing done by family and peers
21. Length of prewriting time at home
22. Need for special writing place
23. Degree of satisfaction with product
24. Sense of completion
25. Sense of importance of writing

Pianko noted that some of these variables had been identified by prior researchers including Emig, Sawkins, Stallard, Graves, Mischel, and Britton, who had studied precollege subjects. She wanted to examine the variables in a college population.

Odell and Goswami identified the variables they wanted to study, based on their analysis of the interviews and on the question they were trying to answer: Are writers in nonacademic settings sensitive to rhetorical issues? Their *independent variables* were

1. Administrators
2. Caseworkers
3. Pink memos
4. White memos
5. Letters

Their *dependent variables* were

6. Audience-based reasons:
   a. Status of audience
   b. Personal knowledge of or relationship with audience
   c. Personal characteristic of audience
   d. Anticipated or desired action on the part of the audience
7. Writer-based reasons:
   a. Writer's role or position in the organization
   b. Ethos or attitude the writer wishes to project or avoid
   c. Writer's feelings about the subject or task at hand
8. Subject-based decisions:
   a. Importance of the topic dealt with in the writing
   b. Desire to provide an accurate, complete, nonredundant account
   c. Desire to document a conclusion the writer has drawn
9. Linguistic features:
   a. T-units
   b. T-unit length
   c. Clause length
   d. Clauses per T-unit
   e. Passive constructions per T-unit
   f. Cohesive markers used

They cited as bases for variables 6–8, theoretical distinctions introduced by Gibson, Kinneavy, and Halliday (p. 208), and as bases for the linguistic variables, the research of Crowhurst and Watson and the theories of Gibson, Hake and Williams, and Halliday and Hasan (p. 210). Notice that Odell and Goswami cluster their variables into larger constructs: audience-based, writer-based, subject-based, and linguistic-based. Such clustering enables them to achieve better reliability among coders and better validity because theoretical bases can more readily be cited for these larger constructs.

Diederich's *independent variables* were

1. Six occupational fields of the judges
2. 53 judges

The *dependent variables* were

3. Ideas
4. Organization
5. Wording
6. Flavor
7. Mechanics (usage, punctuation, spelling, handwriting)

He gleaned the variables in part from the comments made by the 53 judges. He selected a range of educated readers as variables because he wanted to study reliability, or repeatability, of grading (see Appendix I).

Hypotheses

One key to good quantitative descriptive research is the use of alternative hypotheses that pose different explanations for whatever phenomena are under study. Guided by these hypotheses, researchers measure variables and use them to test the alternative hypotheses. Odell and Goswami had alternative hypotheses: that choices made by the writers were rhetorically based (concern for context—speaker, subject, and audience); or that the choices were a-rhetorical ones. This latter hypothesis was defined only as a "rule that he or she follows in all circumstances" (p. 211). Their analysis showed that "caseworkers and administrators rarely justified their preference by referring to a-rhetorical rules" (p. 213). These rare instances were not exemplified. When a quantitative descriptive study identifies alternative hypotheses and relates variables to them, its explanations and generalizations have more power.

Statistical analyses for two variables

In quantitative descriptive research, a first purpose of statistical analysis is to test relationships among two variables at a time in *one group of subjects* by observing, measuring, and interrelating them to find out if they are related or independent of each other. A second purpose of statistics is to show differences among *two or more groups of subjects* on one variable at a time. For example, when two or more groups of subjects are known to differ on one variable such as sex or grade, the researcher asks whether they also differ on such variables as syntactic fluency and level of writing anxiety.

To relate the variables, a researcher can use a variety of statistical analyses, the choice being determined by the type of variable and the purpose of the analysis (see Figure 5-1).

1. To relate *nominal* variables with other *nominal* variables within a group, a researcher can use
   a. Percentages, proportions,
   b. Counts, frequencies, enumerations,
   c. Chi Square (χ^2),
   d. Phi (ϕ) Coefficient (for dichotomous variables).

Data Type	Differences Between Two or More Groups	Relationships Among Variables for One Group
Nominal	Chi Square, χ^2	Chi Square, χ^2 Phi Coefficient
Interval	Analysis of Variance using F test *t* test Point Biserial Correlation	Correlation
Rank Order	Wilcoxon, T Mann–Whitney, U	Spearman's rho

Figure 5-1 Statistical analyses for two variables.

2. To relate *interval* with *interval* variables in one group, a researcher can use correlational analysis.
3. To relate *interval* variables to *nominal* variables among groups, a researcher can use
   a. Analysis of variance and *F*-test,
   b. *t*-Test (for two groups),
   c. Point biserial correlation (for two groups).
4. To relate *nominal* variables to *rank-ordered* variables, a researcher can use
   a. Wilcoxon T (two groups),
   b. Mann–Whitney *U* (two groups).
5. To relate *rank-ordered* variables to other *rank-ordered* variables in one group, a researcher can use Spearman's rho (ρ). (This is the rank order variable analogue to correlational analysis.)

MEAN

For both nominal and interval variables, the researcher can determine the means. As explained in Appendix I, the *mean* is the sum of all scores for a variable, divided by the number of subjects. Pianko calculated the means of interval variables like overall composing time and composing rate per minute. Odell and Goswami calculated the means for frequency variables such as T-units, clauses per T-unit, passive constructions per T-unit, and cohesive markers per T-unit. When there are frequency variables, the means can be percentages or proportions of the group, for example, percentage or proportion of cohesive markers per T-unit.

VARIANCE AND STANDARD DEVIATION

The measure of individual differences or dispersion of scores in each variable is indicated statistically by the *variance* and *standard deviation* (see Appendix I). The variance is an indication of the amount of individual differences in a variable. An important goal in some research is to determine the percentage of variance of the dependent variable that is explained by the independent variables—in other words, to establish the amount of individual difference that can be accounted for as a result of the relationship of one variable to another.

The standard deviation is the square root of the variance. In a normal curve, for large samples there are about six standard deviations in a set of scores: three up from the mean and three down from the mean for any variable. In Chapter 6 of his report, Diederich presents the standard deviation for writing test essays, comparing it with percentiles, standard scores, range of scores, and letter grades (pp. 24–32; and see Figure 5-2).

For smaller samples there are only about five standard deviations in the range of scores on a variable. (For large samples, it is more likely that the rarer scores, extremely high and low ones, will appear. Hence the larger the range will be.) The standard deviation is widely used in statistical analysis and in psychological measurement.

ANAYLSES FOR NOMINAL WITH NOMINAL VARIABLES

Frequencies, Proportions, and Percentages

Analyses often show frequency counts—the number of items, people, units—that belong in any nominal category. Counts can be turned into

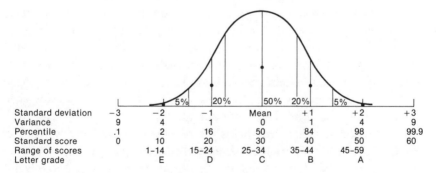

		5%	20%	50%	20%	5%		
Standard deviation	−3	−2	−1	Mean	+1	+2	+3
Variance	9	4	1	0	1	4	9
Percentile	.1	2	16	50	84	98	99.9
Standard score	0	10	20	30	40	50	60
Range of scores		1–14	15–24	25–34	35–44	45–59	
Letter grade		E	D	C	B	A	

Figure 5-2 Normal curve for standard deviation and other scores. ● Indicates average paper in each interval; % in each interval. *Source:* Diederich (1974), p. 24. Copyright © 1974 by the National Council of Teachers of English. Reprinted by permission of the publisher.

proportions by dividing the number in a group by the total number. These counts can also be turned readily into *percentages* by dividing the number in the category by the total number of subjects and then multiplying by 100. Most of Pianko's analyses were displayed either in tables showing frequencies and percentages or in her descriptive text that cites percentages. She made little attempt to interrelate statistically her 22 variables. This was just as well because she only had data for 17 subjects. To have thus cross-classified the data would have reduced the number in any cell to very few, making a discernible trend difficult to observe. Odell and Goswami also reported the number and percentage of types of reasons given for choices in writing.

Chi Square

Chi-Square analysis, like all statistical analyses of nominal variables, is based on the number of subjects or objects that are found in given categories: number of women, number of people 21 years old, number of papers with fragments, etc. It shows either the possible relationship between two or more nominal variables, or the possible differences between two or more groups of subjects (the groups being nominal variables) on one nominal variable at a time. (See Appendix I for more on Chi Square.) Pianko used Chi Square to detect differences between nominal variables like "male/female," "remedial/traditional," and nominal variables like "planning behavior and writer's concerns." (See her variable list, numbers 9–14.) Table 5-1 shows the results of this analysis.

ANALYSES FOR INTERVAL WITH INTERVAL DATA

Remember that *interval* data are ordered, equal units, coming from sources like test scores, ratings, grades, or timing. In the Pianko study, the interval data included prewriting time, composing time, rate of composing, and rereading time. These data, coming from the observers' ratings, test scores, self-rating, checklists, etc., are said to be *intervally measured data*. When they are going to be analyzed, they can be put into a data matrix with the variables' scores or ratings (for each person or composition) listed horizontally, and the persons or compositions vertically. This matrix helps the researcher to see quickly whether the data on all the K variables have been collected for all subjects. Analyses can then be done to determine the average of each variable for this group of subjects (the mean) and the individual differences among the subjects on each variable (the standard deviation) and the relation-

Table 5-1 χ^2 Analysis of Dependent Variables Nine through Fourteen for Age Groups

Factors	Male[a] Frequency of Response	Percent	Female[b] Frequency of Response	Percent	χ^2 value[c]
9. Planning behavior—mental	7	77.8	7	87.5	. . .[c]
10. Attitude toward the writing —positive	4	44.4	5	62.5	. . .[c]
11. Stylistic concerns—none	5	55.6	3	37.5	1.647
12. Consideration of purpose— none	6	66.7	4	50.0	1.316
13. Knowledge of ideas prior to writing—none	5	55.6	3	37.5	
14. Writer's concerns:					2.034
getting ideas across	5	55.6	3	37.5	
mechanics and usage	1	11.1	3	37.5	
correct word	2	22.2	2	25.0	
none	1	11.1	0	0.0	

Source: Pianko (1981), Table 4. Copyright © 1981 by the National Council of Teachers of English. Reprinted by permission of the publisher.

[a] There were nine students in the male group.

[b] There were eight students in the female group.

[c] For a 2×2 table when the marginal values are small the appropriate procedure to use is Fisher's Exact Test rather than χ^2. Neither Fisher's Exact Test was significant.

ship among all variables. A type of analysis that can be done on interval variables is discussed below.

Correlation

A common analysis done on interval data is *correlation*, which measures the relationship between pairs of interval variables. The correlation is an index showing the strength of the relationship of the scores on two variables. A straight-line (linear) relationship between two variables yields the minimum correlation between them. But if the true relationship between the two is a curve, that fact will be more accurately shown by a nonlinear correlation that will show a higher relationship than a simple linear correlation.

Pianko's analytical methods were appropriate for the kinds of variables she wanted to correlate. Because her 22 quantitative variables were either interval or dichotomized nominal variables, her analysis could have been entirely correlational, simply one pass through the computer, yielding the means and standard deviations of all 22 variables and a correlational matrix of 25 × 25 variables (22 dependent and 3 independent variables). This correlational matrix would have yielded a complete picture of all the relationships among the 25 quantitative descriptive variables. But clearly there would have been hazards in this type of analysis with only 17 subjects. If all 300 possible nonredundant correlations had been calculated, 15 (5%) would have been expected to be "significant" by chance alone even if there were no real patterns of relationship among the 25 variables.

Today all correlations are normally done by a statistical program on a computer. The advantage of correlations is that a researcher can present a great deal of information about the subjects succinctly, even though some information is lost.

ANALYSES FOR NOMINAL WITH INTERVAL VARIABLES

Analysis of Variance and the F-Test

Analysis of variance is used when one of the variables is nominal (e.g., mental planning behavior, positive attitude toward writing, or sex) and the other is interval (e.g., prewriting time, number of revisions per 100, grade on a composition). (See Appendix I for more explanation.) Pianko used analysis of variance to check the relationship between the interval variable, "prewriting time," and the nominal variable, "class status—remedial/traditional." She found a significant difference between the two groups in relation to prewriting time, calculating an F-ratio of 5.26. She did not find a significant difference in prewriting based on her other nominal categories—age or sex. Table 5-2 shows the results for Pianko's other seven factors based on class status groups.

Odell and Goswami's use of analysis of variance was appropriate, helping them to point out that the writers in their study were sensitive to rhetorical context in making their choices. They decided in advance of any analysis, before examining the data, which of the many possible comparisons of averages they would make.

T-*Test*

The *t-test* is often used when one variable is dichotomous nominal and the other one is interval. (See Appendix I for further discussion.) When Pianko, for example, wanted to show differences between two levels of the dichotomous nominal variable, "instructional class type (traditional type vs. remedial)" for an interval variable, "composing time," she could have used the *t*-test.

RANK ORDER STATISTICS

Rank order statistics, used with rank order data, can be employed to show relationships among variables or to show differences between

Table 5-2   Analysis of Variance of Dependent Variables One through Eight for Class Status Groups

	Mean scores		
Factors	Remedial	Traditional	F-ratio
1. Prewriting time (minutes)	1.00	1.64	5.2552[a]
2. Composing time (minutes)	35.75	43.29	1.1831
3. Rate of composing (words/minute)	9.31	9.29	0.0001
4. Rereading time (minutes)	3.20	3.71	0.0741
5. Revisions	2.56	3.71	1.1543
6. Pauses	11.40	23.43	10.3418[b]
7. Drafts	0.10	0.14	0.0649
8. Rescannings	3.70	11.71	38.3186[b]

Source: Pianko (1981), Table 3. Copyright © 1981 by the National Council of Teachers of English. Reprinted by permission of the publisher.

DF: 1,15
F-ratio required at .05 level—4.54.
F-ratio required at .01 level—8.68.

[a]Significant at or beyond .05 level.
[b]Significant at or beyond .01 level.

and among groups of people on one variable at a time. These rank order statistics form perhaps half of what are called nonparametric statistics. (See Appendix I.)

Multivariate analyses

Each of the above analyses are used to relate one variable with a second variable. *Multivariate analyses* are more complex statistical methods being used today to analyze the relationships among a number of variables. These kinds of analyses are highly appropriate for composition research because of the complex nature of the writing process, the acquisition of literacy, and its contextual environments. Seldom does one variable account fully for another or influence other variables in isolation. To analyze this kind of interdependency and confluence of relationships, researchers use various kinds of multivariate analyses: multivariate analysis of variance, canonical correlation, discriminant analysis, multiple regression analysis, and factor analysis. (See Appendix I for fuller treatments of these statistical analyses.) If researchers want, for example, to study the interrelated influence of several independent variables like sex, mode of discourse, and age, on a dependent variable such as quality of writing, they can use a multivariate analysis. A multiple-regression analysis will help them determine how much of the variance of the dependent variable is accounted for by the independent variables. Multivariate analyses can be used for both quantitative descriptive studies and for experimental research. Let us examine one type of multivariate analysis—factor analysis.

FACTOR ANALYSIS

The purpose of factor analysis is to make larger, overarching comprehensive variables out of a set of smaller individual ones that correlate with one another. By factor analysis, a researcher can use a few larger variables to portray more effectively the major dimensions of a group, a set of papers, a cluster of evaluative attitudes, and so forth. After a researcher has clustered these smaller variables into larger ones, these larger variables are relatively independent of one another.

Diederich first used correlation to find out how the ratings of each reader related to the grades of all other readers. He created a large correlational matrix (53 × 53 variables). He did not report all the results of his correlational analysis, except for the average correlation of .31 because he did a factor analysis, which reduced the relationships found in the 53 × 53 matrix to a smaller set of overarching variables. He found these factors by picking out clusters of variables, in this case readers who were similar to one another but relatively independent of all other readers. He elaborated on these factors, using the qualitative data in the study—the readers' comments about what they liked or disliked in their reading of the papers. The overarching factors he found were ideas, organization, wording, flavor, and mechanics (usage, punctuation, spelling, and handwriting). These factors helped to explain the correlations found in the matrix.

Results

The conclusions of quantitative descriptive research indicate the calculated strengths or weaknesses of relationships among variables. Researchers can also determine significance and the amount of variance explained.

SIGNIFICANCE

The researcher wants to establish that a relationship between two variables is *significant*, in other words, that it is unlikely to have happened by chance alone. In many cases of statistical analysis, the calculated statistic is compared with standard, tabled values of chance distribution of that statistic. If the calculated value of the score in a study is larger than the tabled value, the result is called statistically *significant*, meaning that the strength of the relationship or difference between two variables observed in one group is unlikely to have occurred simply by random chance alone. The *significance* is declared as "*p*" (probability level), usually at the 5% or 1% level, indicating that the result could have been by chance in either 1 out of 20 cases or 1 out of 100 cases. The 1% level suggests a more unlikely result by chance alone. In more modern statistical analyses where computers are used, the value of the statistic is printed with the exact probability value alongside of it. Examining tabled values is thus unnecessary. The most complete reporting practice when using a computer is to report the statistical value and its exact *p* value, and declare the statistical result as "significant" or "nonsignificant" in accordance with one's preestablished acceptable levels of probability, usually the 5% or 1% levels of *p*.

No researcher can answer all questions on all variables about a population. Even when questions are not answered, however, a piece of research makes a contribution by raising good speculations about how people write and about the effect of their writing on other people. Pianko found significance only in the relationship between class status (remedial versus traditional) and prewriting time, number of pauses and rescannings during composing, and stylistic considerations. She suggested that these significances demonstrated a marked difference in the approach students in each category took toward writing: her traditional writers were more conscious of the elements necessary for a well-developed composition—style, purpose, and getting ideas across. They spent more time prewriting, pausing, and rescanning. Her remedial writers, on the other hand, glanced around the room during their pauses, sometimes as a diversion, at other times as a way of finding the correct spelling, the correct word, or something to say next. They usually did not look to their own texts for answers. They did not see writing as playing an important role in their lives; they had little commitment to it, carrying it out quickly and superficially to meet a requirement.

These trends may form hypotheses for other studies. Research with people is essentially attempting to discern patterns of relationships against a background of a great deal of random noise. When samples are small and measurement instruments brief, traditional levels of statistical significance (i.e., the 5% and 1% levels) will be difficult to obtain.

If Pianko had set a 10% or even 20% p level, instead of 5% or 1%, her percentage of significant results would have been much higher. As she concluded, clearly the importance of her work rested on the suggestions it gave about important relationships that needed to be further studied with larger numbers of subjects.

Odell and Goswami tried to answer important questions about the bases for choices in nonacademic writing. Their small sample of subjects allowed them to collect a considerable amount of data with which they could do detailed analyses. The size of their study also allowed them to test interview sheets and to work out a three-point scale for adequacy of product. Their conclusions, on the basis of interviews and content analyses of the writing, identified some reasons for choice in nonacademic writing, thus establishing variables that could be further tested. They were careful not to generalize beyond their subjects. In order to gain more power in their correlations, however, they needed more subjects—at least ten times the variables. In their analysis, they set the traditional 5% and 1% levels of significance. Given their table with 50 tests of statistical significance (see Table 5-3), at the 5% level only about 2.5 of these tests would be significant by chance alone. In fact, 24 were—which suggests that indeed there were true relationships detected among the variables.

AMOUNT OF VARIANCE EXPLAINED

Many researchers want to know more than whether a relationship among variables is significant. They want to calculate the *strength* of the relationship in order to determine the amount of variance in a dependent variable that is explained by a given independent variable or a set of interrelated variables. The greater the differences between groups on any dependent variable, the more an independent variable can be said to have accounted for that difference. In the Pianko study, for example, the greater the variance on her dependent variable, "number of pauses," the greater the percentage of influence on that variance of the independent variable, "traditional type of student," which had an F ratio significant at the p<.01 level. If an independent variable does account for a large percentage of the variance on a dependent measure, relative to the total variance, that variable can be said to have a large influence or a strong relationship to a dependent variable. Pianko, for example, found a significant difference in prewriting time between her remedial and traditional groups. She could have gone on to calculate the extent of that difference compared to the individual differences and to determine the percentage of the variance explained by the independent variable, class type—traditional or remedial.

Two kinds of variance exist: systematic and random. Systematic is the kind of variance in a dependent variable that is explained by the influence of an independent variable. Random variance is what is left unexplained, or assumed to have occurred by chance. The researcher wants to reduce that random variance by trying to find variables that will explain it. One reason why some variance is not accounted for by certain measures is that they themselves are not highly reliable, a problem that will be explained in Chapter 7.

INTERACTIONS

One of the difficulties in forming conclusions comes from the presence of interactions. An *interaction* is the combined effect of one or more variables on another variable. An analysis of variance table identifies such effects. If an interaction is significant, the relationship between two or more variables is larger than or smaller than what might ordinarily be expected by simply adding the effects of one variable to the other. Significant interactions make it difficult for a researcher to generalize about the relationships between the independent variables and any dependent variable.

Odell and Goswami did an analysis of variance (writer × product type) for the following dependent variables: mean number of T-units, mean T-unit length, mean clause length, mean number of clauses per T-unit, mean number of between T-unit coordinate conjunctions, mean number of passives per T-unit, and mean number of reference, substitution, conjunction, and lexical cohesive markers per T-unit. They found that the writer variable was significant for seven of the ten dependent variables. But there were significant interactions between writer and product type on four of the dependent variables.

In an interaction, one of the independent variables may be called a moderator variable. A moderator variable (e.g., sex of writer) influences the relationship between an independent variable like type of writing (letter or memo) and a dependent variable like T-unit length. If males produce memos with longer T-units while females produce letters with longer T-units, then one can say that there is an interaction between sex and type of writing as measured by T-unit length. While interactions and moderator variables can be systematically examined through analysis, the concept of interaction is more general, applying whenever a researcher discovers that a variable has a differing influence on the results in the context of a second, third, or fourth variable (see Figure 5-3).

Not all of Odell and Goswami's writers produced sentences of proportionately different T-unit lengths, clauses per T-unit, passives per

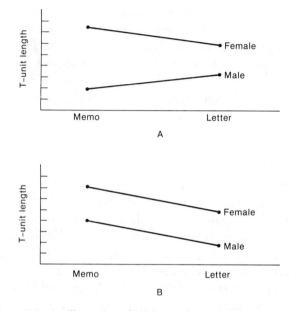

Figure 5-3 An illustration of (A) interaction and (B) no interaction.

Table 5-3 Analysis of Variance in Administrators' Writing

Variable	Writer (4,235)	Type (2,235)	Interaction of Writer and Type (8,235)	Pink vs. White Memos (1,235)	Memos vs. Letters (1,235)
Number of T-Units	7.446**	8.613**	1.801	15.37**	1.85
T-Unit Length	8.491**	5.011**	3.164**	7.84**	2.17
Clause Length	5.107**	1.022	1.682	.85	1.19
Clauses Per T-unit	18.539**	.983	2.041*	1.76	.59
Between T-unit Coordination	1.715	1.399	1.081	.71	2.14
Passives Per T-unit	2.477*	8.728**	3.832**	16.41**	2.07
Reference Cohesive Ties Per T-unit	5.760**	8.200**	1.369	13.25**	2.87
Substitution Cohesive Ties Per T-unit	5.610**	3.949**	.937	.07	8.84**
Conjunction Cohesive Ties Per T-unit	1.502	3.475	1.508	6.43*	2.12
Lexical Cohesive Ties Per T-unit	17.511	26.607**	2.728**	51.88**	1.12

Source: Odell and Goswami (1982), Table 3. Copyright © 1982 by the National Council of Teachers of English. Reprinted by permission of the publisher.

*indicates significance at the .05 level
**indicates significance at the .01 level

T-unit, and lexical cohesive ties per T-unit over the three types of writing they did. Interactions made it impossible to generalize about product types over writers or writers over product types (see Table 5-3). Had there been no significant interactions, the number of clause lengths and T-units might be the same or different among writers or among the three product types, but with differences across writers by the three types of writing, these differences were proportional across writers. In other words, with no interactions, knowledge of T-unit length in one type of writing allowed them to predict the T-unit lengths in the other two types of writing. With significant interactions, these T-unit lengths no longer were proportionate across writing types and could not be predicted. See Figure 5-4 for the difference between an interaction and no interaction.

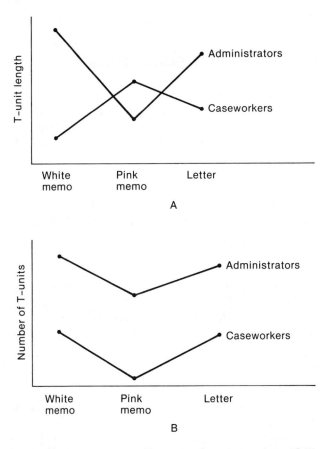

Figure 5-4 Hypothetical illustration of (A) a significant interaction and (B) a nonsignificant interaction.

Summary

Quantitative research is helpful in systematically determining relation-ships among variables in a large group of subjects. Researchers use a variety of statistical analyses to determine relationships among vari-ables. Common analyses include Chi Square and Phi Coefficient for nominal variables, correlation for interval variables, analysis of vari-ance and *t*-tests for relationships between nominal and interval vari-ables, and multivariate analysis for interrelationships among several variables. Investigators try to establish significant relationships, to de-termine the amount of variance in dependent variables that indepen-dent variables explain, and to detect interactions. By discriminating be-tween more and less important variables, quantitative descriptive research helps build theories about the composing processes, the con-texts of writing, and the pedagogy of writing, but it is usually inade-quate to show cause-and-effect relationships among variables.

Checklist for reading "Quantitative Descriptive Studies"

1. What were the questions or hypotheses posed for the study?
2. What importance did the literature review give to the study?
3. What are the variables? Were they nominal or interval? What were the sources of these variables?
4. Who were the subjects? What ratio existed between subjects and variables?
5. What were the data collection procedures?
6. What were the methods of statistical analysis? Were they appro-priate for the kinds of variables being studied?
7. What were the conclusions? Were there any interactions that af-fected these conclusions?
8. Were any relationships significant? At what level?
9. What percentage of the variances of the dependent measures was explained by the independent variables?

References

Research Design and References

Amick, D.J., & Walberg, H.J. (Eds.) (1975). *Introductory multivariate analysis*. Berkeley, Calif.: McCutchan.

Asher, J.W. (1976). *Educational research and evaluation methods*. Boston: Little, Brown.

Cohen, J.A. (1977). *Statistical power analysis for the behavioral sciences*. New York: Academic Press.

Edwards, A.L. (1985). *Experimental design in psychological research.* 5th edition. New York: Harper & Row.

Finn, J.D. (1974). *A general model for multivariate analysis.* New York: Holt, Rinehart & Winston.

Hummel, T.J., & Sligo, J.R. (1971). Empirical comparison of univariate and multivariate analysis of variance procedures. *Psychological Bulletin, 76,* 49–57.

Kaiser, H.F. (1958). The varimax criterion for analytic rotation in factor analysis. *Psychometrika, 23,* 187–200.

Kaiser, H.F. (1960). The application of electronic computers to factor analysis. *Educational Psychological Measurement, 20* 141–151.

Kerlinger, F.N. (1986). *Foundations of behavioral research,* 3rd edition. New York: Holt, Rinehart & Winston.

Kerlinger, F.N., & Pedhazur, E.J. (1973). *Multiple regression in behavioral research.* New York: Holt, Rinehart & Winston.

Smith, J.K. (1983). Quantitative versus qualitative research: An attempt to clarify the issue. *Educational Researcher, 12,* 6–13.

Timm, N.H. (1975). *Multivariate analysis with applications in education and psychology.* Monterey, Calif.: Brooks/Cole.

Vockell, E.L. (1983). *Educational research.* New York: Macmillan.

Walker, J.T. (1984). *Statistics the way it should be.* London: Sage Publications.

Winer, B.J. (1962). *Statistical principles in experimental research design.* New York: McGraw-Hill.

Quantitative Descriptive Studies

Afghari, A. (1984). Grammatical errors and syntactic complexity in three discourse categories: An analysis of English compositions written by Persian speakers. *DAI, 45,* 03A.

Allred, W.G. (1984). A comparison of syntax in the written expression of learning disabled and normal children. *DAI, 45,* 05A.

Bamberg, B. (1984). Assessing coherence: A reanalysis of essays written for the National Assessment of Educational Progress, 1969–1979. *RTE, 18*(3), 305–319.

Barnes, J.A. (1983). An analysis and comparison of reading and writing levels of third, fifth, and seventh grade students as measured by readability formulas. *DAI, 45,* 01A.

Benton, S.J., Kraft, R.G., & Glover, J.A. (1984). Cognitive capacity differences among writers. *Journal of Educational Psychology, 76,*(5), 820–834.

Boice, R. (1985). Cognitive components of blocking. *Written Communication, 2*(1), 91–104.

Bond, S., & Hayes, J. (1984). Cues people use to paragraph texts. *RTE, 18,* 147–167.

Bridwell, L.S. (1980). Revising strategies in twelfth grade students' transactional writing. *RTE, 14,* 197–222.

Bridwell, L.S. (1982). Revising processes in twelfth grade students' transactional writing. *DAI, 40,* 11A.

Bryant, B.R. (1984). The relationship of holistic and objective measurement of transcription and composition components in written language. *DAI, 45,* 08A.

Burleson, B.R., & Rowan, K.E. (1985). Are social-cognitive ability and narrative writing skill related? *Written Communication, 2*(1), 25–43.

Burtoff, M.J. (1984). The logical organization of written expository discourse in English: A comparative study of Japanese, Arabic, and native speaker strategies. *DAI, 45,* 09A.

Carpenter, C. (1984). Relationships between two distinct personality types, and their composing processes. *DAI, 45*, 03A.

Chesky, J.A. (1984). The effects of prior knowledge and audience on writing. *DAI, 45*, 09A.

Christensen, M. (1977). The college-level examination program's freshman English equivalency examinations. *RTE, 11*, 186–192.

Ciani, A. (1976). Syntactic maturity and vocabulary diversity in the oral language of first, second, and third grade students. *RTE, 10*, 150–156.

Clark, C.L. (1984). Reader, writer, and text context variables as influences on the comprehensibility of text. *DAI, 45*, 03A.

Collins, J., & Williamson, M. (1981). Spoken language and semantic abbreviations in writing. *RTE, 15*, 5–22.

Collins, J., & Williamson, M. (1984). Assigned rhetorical context and semantic abbreviation. In R. Beach and L. Bridwell (Eds.). *New directions in composition research.* New York: Guilford Press.

Cooper, C., & Michalak, D. (1981). A note on determining response styles in research on response to literature. *RTE, 15*, 163–169.

Cox, B., & Sulxby, E. (1984). Children's use of reference in told, dictated, and handwritten stories. (RTE, 18(4), 345–365.

Daiute, C. (1984). Performance limits on writers. In R. Beach and L. Bridwell (Eds.). *New directions in composition research.* New York: Guilford Press.

Daly, J. (1979). Writing apprehension in the classroom: Teacher role expectancies of the apprehensive writer. *RTE, 13*, 37–44.

Daly, J., & Hailey, J.L. (1984). Putting the situation into writing research: State and disposition as parameters of writing apprehension. In R. Beach and L. Bridwell (Eds.). *New directions in composition research.* New York: Guilford Press.

Daly, J., & Shamo, W. (1978). Academic decisions as a function of writing apprehension. *RTE, 12*, 119–126.

Davis, D. (1977). Language and social class: Conflict with established theory. *RTE, 11*, 207–218.

Diederich, P. (1974). *Measuring growth in English.* Champaign, Ill.: National Council of Teachers of English.

Dillon, D. (1978). Semantic development of selected lexical items as studied through the process of equivalence formation. *RTE, 12*, 7–20.

Dilworth, C., Reising, R., & Wolfe, D. (1978). Language structure and thought in written composition: Certain relationships. *RTE, 12*, 97–106.

Domaracki, J. (1984). The relationship of the amount and structure of children's prior knowledge to reading comprehension of expository prose. *DAI, 45*, 03A.

Fagan, W. (1982). The relationship of the "maze" to language planning and production. *RTE, 16*, 85–96.

Ferguson, D.L. (1984). The effect of audience on communicative competence in oral and written expression of learning disabled students at two age levels. *DAI, 45*, 12A.

Finnell, L.D. (1984). The comprehension of restrictive and nonrestrictive relative clause information in written language by seventh grade high and average reading achievers in a suburban junior high school level. *DAI, 45*, 08A.

Foster, D. (1984). *Coherence, cohesion, and deixis.* ERIC Document Reproduction Service No. ED 245 223.

Fox, B. (1983). *Linguistic correlates of two writing functions, at two grade levels, and two achievement levels.* ERIC Document No. ED 240 572.

Fox, R. (1984). A study of metaphor in the writing of nine- and thirteen-year-olds, college freshman, and graduate students in the humanities and in the sciences. *DAI, 45,* 06A.

Freedman, S. (1981). Influences on evaluators of expository essays: Beyond the text. *RTE, 15,* 245–255.

Gaines, B.E. (1983). An analysis of the relationship between teaching behavior and student achievement in learning to write essays about poetry. *DAI, 45,* 03A.

Garner, R., Belcher, V. Winfield, E., & Smith, T. (1985). Multiple measures of text summarization proficiency: What fifth-graders do? *RTE, 19*(2), 14–153.

Gebhard, A. (1978). Writing quality and syntax: A transformational analysis of three prose samples. *RTE, 12,* 211–232.

Geilker, N.F. (1984). A comparison of professionals' responses to selected errors of usage and items of disputed usage in formal written English. *DAI, 45,* 04A.

Goodman, K., & Bridges Bird, L. (1984). On the wording of texts: A study of intra-text word frequency. *RTE, 18,* 119–146.

Gourley, J.W. (1984). Discourse structure: Expectations of beginning readers and readability of text. *Journal of Reading Behavior, 16*(3), 169–188.

Gowda, N.S. (1984). An exploration of the role of language awareness in high school students' reading and writing. *DAI, 45,* 08A.

Graesser, A., Hopkinson, P., Lewis, E., & Bruflodt, H. (1984). The impact of different information sources on idea generation. *Written Communication, 1,* 341–364.

Green, M. (1985). Talk and Doubletalk: The development of metacommunication knowledge about oral language. *RTE, 19*(1), 9–24.

Griffiths, A.H. (1984). Prospective secondary teachers' responses to student use of Black English in written compositions. *DAI, 45,* 12A.

Groff, P. (1978). Children's spelling of features of black English. *RTE, 12,* 21–28.

Grubb, M.H. (1983). The writing proficiency of selected ESL and monolingual English writers at three grade levels. *DAI, 45,* 01A.

Harris, W. (1977). Teacher response to student writing: A study of the response patterns of high school English teachers to determine the basis for teacher judgment of student writing. *RTE, 11,* 175–185.

Hayes, J.R., & Flower, L.S. (1984). *A cognitive model of the writing process of adults.* ERIC Document Reproduction Service No. ED 240 608.

Higgs, R.O. (1984). The impact of discourse mode, syntactic complexity, and story grammar on the writing of sixth graders. *DAI, 45,* 01A.

Hogan, C. (1977). Let's not scrap the impromptu test essay yet. *RTE, 11,* 219–226.

Hrach, O.E. (1983). The influence of rater characteristics on composition evaluation practices. *DAI, 45,* 02A.

Hull, G.A. (1983). The editing process in writing: A performance study of experts and novices. *DAI, 45,* 03A.

Hunt, K.W. (1965). *Grammatical structures written at three grade levels.* Champaign, Ill.: National Council of Teachers of English.

Jamieson, B.C. (1984). Features of the thematic and information structures of the oral and written language of good and poor writers. *DAI, 45,* 08A.

Kagan, D. (1980). Run-on and fragment sentences: An error analysis. *RTE, 14,* 127–138.

Kane, R. (1983). A longitudinal analysis of primary children's written language in relation to reading comprehension. *DAI, 45,* 03A.

Keller, T.A. (1984). A study of the relationship of obscuring writing posture and reading disability. *DAI, 45,* 10A.

Kroll, B.M. (1984). Audience adaptation in children's letters. *Written Communication, 1*(4), 407–427.

Kroll, B.M. (1985). Rewriting as complex story for a young reader: The development of audience-adaptive writing skills. *RTE, 19*(2), 120–139.

Kucer, S.B. (1983). *Controlling the writing process: Not a monolithic process.* ERIC Document Reproduction Service No. ED 240 615.

Langer, J. (1984). The effects of available information on responses to school writing tasks. *RTE, 18,* 27–44.

Ledford, S.Y. (1984). The relationship among selected sixth-grade students' reading schemata, reading achievement, and writing sophistication. *DAI, 45,* 03A.

Leong, C.K. (1984). Cognitive processing, language awareness, and reading in grade 2 and grade 4. *Contemporary Educational Psychology, 9,* 369–383.

Liao, T.S. (1984). A study of article errors in the written English of Chinese college students in Taiwan. *DAI, 45,* 07A.

Lin, J.C. (1984). Chinese students' English compositions in terms of the nonlinguistic variable of sociocultural difference: A contrastic rhetoric. *DAI, 45,* 08A.

Maimon, E., & Nodine, B. (1978). Measuring syntactic growth: Errors and expectations in sentence-combining practice with college freshmen. *RTE, 12,* 233–244.

Matsuhashi, A. (1981). Pausing and planning: The tempo of written discourse production. *RTE, 15,* 113–134.

Matsuhashi, A., & Quinn, K. (1984). Cognitive questions from discourse analysis. *Written Communication, 1,* 307–340.

McQuillan, M.K. (1984). The measurement of writing abilities in grades 7, 8, and 9. *DAI, 45,* 04A.

Meade, R. (1971). The use in writing of textbook methods of paragraph development. *Journal of Educational Research, 65,* 74–76.

Merrill, S. (1986). Audience adaptiveness in job application letters written by college students: An exploratory study. *DAI, 48,* 121A.

Meyers, G.D. (1984). Exploring differences in cognitive strategies among and between freshman English students and instructors: A research model. *Writing Instructor, 3*(2), 61–74.

Mosenthal, P., Davidson-Mosenthal, R., & Krieger, P. (1980). How fourth graders develop points of view in the classroom writing. *RTE, 15,* 197–214.

Newkirk, T. (1984a). Direction and misdirection in peer response. *CCC, 25*(3), 301–311.

Newkirk, T. (1984b). How students read papers: An exploratory study. *Written Communication, 1,* 283–306.

Nold, E., & Freedman, S. (1977). An analysis of readers' responses to essays. *RTE, 11,* 164–174.

Odell, L., & Goswami, D. (1982). Writing in a non-academic setting. *RTE, 16,* 201–223.

Oi, K.M. (1984). Cross-cultural differences in rhetorical patterning: A study of Japanese and English. *DAI, 45,* 08A.

Parla, J. (1984). The written Spanish of Puerto Rican bilinguals in a situation of language contact: An error analytic study. *DAI, 45,* 08A.

Payne, E.M. (1984). A profile of the written vocabulary of the adult basic writer. *DAI, 45,* 04A.

Pellegrini, A.D. (1984). Symbolic functioning and children's early writing: The relations between kindergartners' play and isolated word-writing fluency. In R. Beach & L. Bridwell (Eds.). *New directions in composition research.* New York: Guilford Press.

Pellegrini, A.D., Galda, L., & Rubin, D.L. (1984). Context in text: The development of written language in two genres. *Child Development, 55*, 1549–1555.

Pendarvus, E.D. (1983). A comparison of gifted students' written language with the written language of average students. *DAI, 45*, 01A.

Pfeifer, J. (1983). *What happens to writing apprehension in a reading class?* ERIC Document Reproduction Service No. ED 240 542.

Pianko, S.H. (1978). The composing acts of college freshman writers: A description. *DAI, 38*, 07A.

Pianko, S.H. (1981). A description of the composing processes of college freshmen. *RTE, 13*, 5–22.

Popplewell, S.R. (1984). A comparative study of the writing and reading achievement of children, ages nine and ten, in Great Britain and the United States. *DAI, 45*, 09A.

Powers, W., Cook, J., & Meyer, R. (1979). The effect of compulsory writing on writing apprehension. *RTE, 13*, 225–230.

Prater, D., & William, P. (1983). Effects of modes of discourse on writing performance in grades four and six. *RTE, 17*, 127–134.

Price, G., & Graves, R. (1980). Sex differences in syntax and usage in oral and written communication. *RTE, 14*, 147–153.

Quinn, D.P. (1984). Rhetorical analysis of intellectual processes in student writing: Linguistic cues in the quality-rated writing of college pre-freshman. *DAI, 45*, 09A.

Rafoth, B. (1984). Audience awareness and adaptation in the persuasive writing of proficient and nonproficient college freshmen. *DAI, 45*, 09A.

Rafoth, B., & Combs, W. (1983). Syntactic complexity and readers' perceptions of an author's credibility. *RTE, 17*, 165–168.

Ritter, E. (1979). Social perspective-taking abilities, cognitive complexity and listener-adapted communication in early and late adolescence. *Communication Monographs, 46*, 40–51.

Robinson, S.F. (1984). Coherence in student writing. *DAI, 45*, 06A.

Rodrigues, R. (1980). Bilingual and monolingual English syntax on the Isle of Lewis, Scotland. *RTE, 14*, 139–140.

Rottweiler, G.P. (1984). Systematic cohesion in published academic English: Analysis and register description. *DAI, 45*, 08A.

Rubin, D., & Piché, G. (1979). Development in syntactic and strategic aspects of audience adaptation skills in written persuasive communication. *RTE, 13*, 293–316.

Rubin, H. (1984). An investigation of the development of morphological knowledge and its relationship to early spelling ability. *DAI, 45*, 10A.

Rummell, M.K. & Dykstra, R. (1983). Analogies produced by children in relationship to grade level and linguistic maturity. *RTE, 17*, 51–60.

Sanchez-Escobar, A. (1984). A contrastive study of the rhetorical patterns of English and Spanish. *DAI, 45*, 05A.

Sanderson, H. (1976). Student attitudes and willingness to spend time in unit mastery learning. *RTE, 10*, 191–198.

Schumacher, G., & Martin, D. (1983). *A categorical analysis of writing protocols of English school children.* ERIC Document Reproduction Service No. ED 240 578.

Schumacher, G., Klare, G., Cronin, F., & Moses, J. (1984). Cognitive activities of beginning and advanced writers: A pausal analysis. *RTE, 18*, 169–187.

Schuessler, B., Gere, A., & Abbott, R. (1981). The development of scales measuring teacher attitudes toward instruction in written composition: A preliminary investigation. *RTE, 15*, 55–63.

Shanahan, T. (1984). Nature of the reading–writing relationship: An exploratory study. *Journal of Educational Psychology, 76*, 466–477.

Springate, K.W. (1984). Developmental trends and interrelationships among preprimary children's knowledge of writing and reading readiness skills, *DAI, 45*, 09A.

Stahl, A. (1974). Structural analysis of childrens' compositions. *RTE, 8*, 184–205.

Stahl, A. (1977). The structure of children's compositions: Developmental and ethnic differences. *RTE, 11*, 156–163.

Stewart, M. (1978). Syntactic maturity from high school to university: A first look. *RTE, 12*, 37–46.

Stoddard, S.E. (1984). Texture, pattern, and cohesion in written texts: A study with a graphic perspective. *DAI, 45*, 09A.

Swan, B. (1979). Sentence combining in college composition: Interim measures and patterns. *RTE, 13*, 217–224.

Taylor, C.A. (1984). The relative effects of reading or writing a prose or diagrammatic summary upon the comprehension of expository prose. *DAI, 45*, 04A.

Thistlethwaite, L.L. (1983). Effects of using text-structure and self-generated questions on comprehension of information from three levels of text-structure as measured by free recall. *DAI, 45*, 02A.

Thompson, R. (1981). Peer grading: Some promising advantages for composition research and the classroom. *RTE, 15*, 172–174.

Vande Kopple, W. (1983). Something old, something new: Functional sentence perspective. *RTE, 17*, 85–99.

Vargus, N.R. (1984). Socio-cognitive constraints in transaction: Letter writing over time. *DAI, 45*, 09A.

Walker, S. (1984). The identification of defects in legal writing and communication as identified for selected New York City attorneys. *DAI, 45*, 07A.

Washington, U.M. (1984). Teachers' responses to children's use of nonstandard English during reading instruction. *DAI, 45*, 06A.

Wesdorp, H. (1982). *Sco Rapport*. Amsterdam: Stichting Centrum voor Onderwijsonderzoek van de Universiteit van Amsterdam.

Winograd, P.W. (1984). Strategic difficulties in summarizing tests. *Reading Research Quarterly, 19*(4), 404–425.

Witte, S. (1983). Topical structure and invention: An exploratory study. *CCC, 34*, 313–319.

Young, R., & Koen, F. (1973). *The tagmemic discovery procedure: An evaluation of its uses in the teaching of rhetoric*. National Endowment for the Humanities. Grant No. EO-5238-71-116.

Zutell, J. (1979). Spelling strategies of primary school children and their relationship to Piaget's concept of decentration. *RTE, 13*, 69–80.

6

Prediction and
Classification Studies

In the last chapter, we pointed out that quantitative descriptive researchers study writing variables in larger groups of subjects to determine whether any significant (not due to chance) relationships exist among them. This chapter discusses prediction and classification research, which seeks to determine the *strength* of a relationship between several variables and a single criterion. One of the uses of this kind of research is to predict behavior such as future grades, attitudes, productivity, educational status, and occupations. In instructional settings, this ability is of major importance if educators intend to assign students fairly to different kinds of instruction, course levels, or curricula. This chapter examines two designs that can determine the strength of significant relationships: *prediction* research that forecasts an interval variable and *classification* research that forecasts a nominal variable. We call these designs *descriptive* research because although researchers relate variables they do not set control conditions or give treatments.

Three prediction studies

Study 1 Betty Suddarth and S. Edgar Wirt (1974). Predicting course placement using precollege information. *College and University, 49,* 186–194.

Suddarth and Wirt collected data for 5000 beginning freshmen in 1971 at Purdue University to develop a basis for placing future incoming freshmen in levels of English composition, mathematics, and chemistry. Using multiple correlations, they analyzed ten predictor variables to determine which best predicted success in these courses. We will discuss below only their work with English placement.

Study 2   Donna Gorrell (1983). Toward determining a minimal competency entrance examination for freshman composition. *RTE, 17,* 263–274.

When the English department of Illinois State University introduced a minimal competency requirement for admission to English 101, the regular-level composition course, they needed a valid, reliable measure of such competency. Gorrell conducted a prediction study to determine which of several factors was the best predictor of success in English 101.

Study 3   Hunter M. Breland and Robert J. Jones (1984). Perceptions of writing skills. *Written Communication, 1,* 101–119.

Breland and Jones studied the criteria raters use when making holistic judgments of brief impromptu essays. They wanted to determine which of forty potential predictor variables would best predict the holistic scores of impromptu essays of 202 Blacks, 202 Hispanics with English as their best language, 200 Hispanics with English as not their best language, and 202 Whites.

Research Design: Prediction

Subjects

In essence, researchers select a present set of students and study the relationship between a criterion like their grades in a college composition course and a potential set of predictors like high-school grades, SAT scores, and essay scores. The relationships that are found are used to determine which variables (high-school grades, SAT scores, essay scores) will best predict the success of future students in the college composition courses.

Variable identification: predictor and criterion variables

In prediction research, variables are divided into two sets: *predictor* variables, which are available at the time the study is begun; and *criterion* variables, collected at the end of the time period.

Predictor variables are those that researchers believe may be useful in forecasting future performance. Usually they are *interval* variables such as prior educational grade-point average (GPA), grades in prior English courses, prior or current test scores, current interview ratings, current attitude measures. In addition, some dichotomous nominal variables like male/female and college/noncollege graduate can be treated as special interval variables, scored simply as 0 and 1, because they have only two categories. Even the nominal variable, type of high-school English course taken, can be quantified readily by dichotomizing each possible response and assigning the number 1 to a "yes" response, and 0 to a "no" response as long as there are not too many types and the quantified responses are one less than the number of possible responses. The relationship of predictor variable to criterion variable can be depicted graphically, as in Figure 6-1.

Suddarth and Wirt's precollege predictor variables were

1. High-school rank
2. High-school GPA
3. SAT verbal score
4. SAT mathematics score
5. Semesters of foreign language in high school
6. High-school English grade
7. Semesters of math

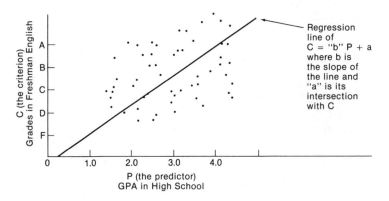

Figure 6-1 Relationship of predictor variable to criterion variable.

8. Semesters of science
9. Math grades
10. Science grades

They also put a nominal variable, sex, into their data matrix. The variable, sex, did not add to their prediction, indicating that there was no discrimination in grades in the university English courses on the basis of sex and that this variable was adequately reflected by prior grades or test scores.

Gorrell's predictor variables were

1. Pre-essay scores
2. Errors on original writing
3. Errors on a revision of O'Donnell's "Aluminum Passage"
4. The cloze test errors
5. ACT English scores
6. ACT social studies scores
7. Reading efficiency
8. Reading comprehension
9. Reading vocabulary

Breland and Jones' predictor variables were

GENERAL CHARACTERISTICS OF DISCOURSE:

1. Statement of thesis and purpose
2. Overall organization
3. Rhetorical strategy
4. Noteworthy ideas: originality of thought or insight
5. Use of supporting materials
6. Writer's tone: voice and attitude
7. Paragraphing and transition
8. Variety of sentence patterns
9. Sentence logic

SYNTACTIC CHARACTERISTICS:

10. Pronoun usage
11. Subject–verb agreement, verb forms
12. Parallel structure
13. Idiomatic usage
14. Punctuation
15. Use of modifiers

LOGICAL CHARACTERISTICS:

16. Appropriate levels of diction
17. Range of vocabulary

18. Precision of diction
19. Use of figurative language
20. Spelling

OTHER VARIABLES:

21. Essay length
22. Number of paragraphs
23. Spelling errors
24. Handwriting
25. Neatness
26. Discourse composite: sum of combined readers' scores for first nine items
27. Syntactic composite: sum of combined readers' scores for characteristics 10–15
28. Lexical composite: the sum of the combined readers' scores for characteristics 16–20
29. Textual composite: the sum of the ratings for characteristics 21–25
30. DSL composite: the sum of all ratings and readers' scores for all 20 characteristics
31–33. Holistic scores
34–40. Test scores: SAT-verbal, vocabulary, reading, SAT-math, TSWE, ECT reported score, ECT objective score

Criterion variables are quantified indicators of success or achievement after a course of instruction. They may be grades in the courses, junior-level competency writing tests, writing on the job five years after graduation, or attitudes toward composing at the end of an instructional period. These criterion variables measure the successfulness of the future behavior for which the predictors are sought. Criterion variables like grades are usually measured as interval variables on a scale of 5, 4, 3, 2, or 1. Suddarth and Wirt's criterion variables were grades in either basic, regular, or advanced composition courses at Purdue and a combined composite. Gorrell's criterion variable was a final holistic rating of an in-class impromptu essay at the end of the semester. Breland and Jones used an ECT holistic score as their criterion variable.

DATA MATRIX

The data for prediction and classification studies are usually placed in a rectangular matrix format. Down the side, *n* represents the number of subjects (people, themes, or paragraphs); across the top, *K* represents the variables of all types in the study (sex, grades, final exam scores, sentence length, ratings, SAT scores). The *K* variables are di-

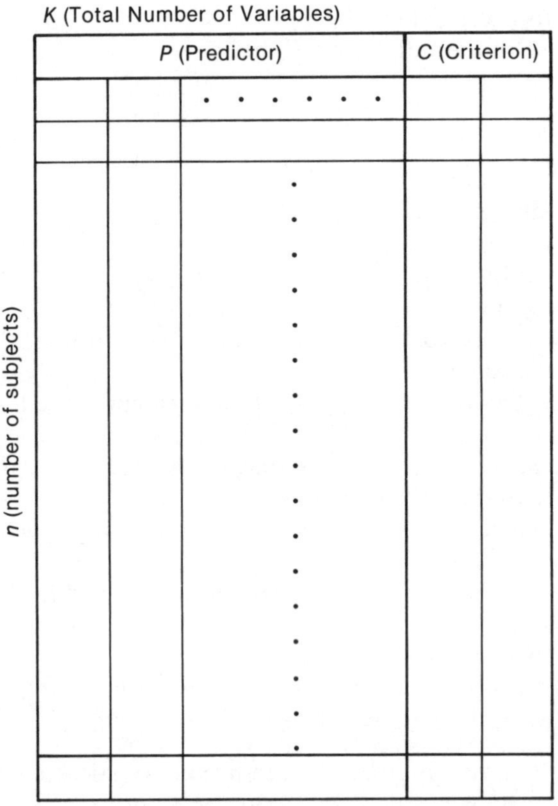

Figure 6-2 $n \times K$ data matrix representation for prediction–classification research.

vided into two sets: predictor *(P)* and criterion *(C)* variables. This data matrix, with one score for each individual for each test, grade, and so forth, is a vertically oriented rectangular matrix because for prediction–classification studies, as well as for quantitative descriptive research, the number of subjects should always be ten times the number of K variables (see Figure 6-2).

CORRELATIONAL MATRIX

Each pair of K variables (across the n subjects) can then be related, producing a matrix of correlations. Because any K variable correlates with itself, this correlation is not recorded. Further, because the correlation between variable 1 and 2 is the same as between variables 2 and 1, and so forth, there is no need to portray half of these correlations in the matrix (see Figures 6-2 and 6-3). These are typically shown as the upper right triangular set of correlations in the matrix.

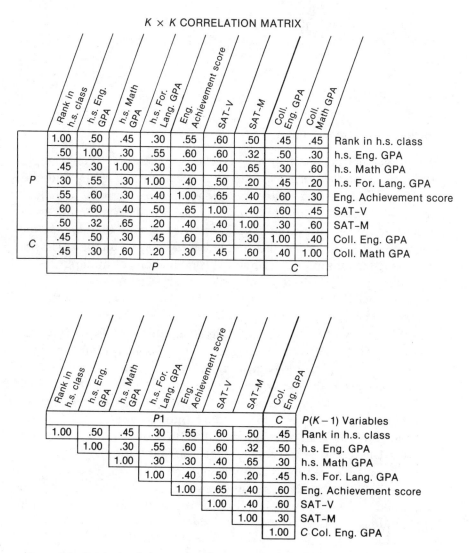

Figure 6-3   Entries in all the matrix cells are correlations. Correlation matrix representations for prediction–classification research.

Suddarth and Wirt at first considered entering data in their data matrix on each of the predictor and criterion variables for each of the 5000 entering students. They eliminated, however, those students with incomplete data (a procedure that is no longer necessary with modern missing-data correlation and regression computing programs), leaving about 2700 students with complete data enrolled in the regular-level freshman English course. They drew a random sample of 675 students, about one-fourth of the regular-level population. (They also drew a sample of 129 students from the basic-level course and 658 students from the advanced English course.) Their total sample was 1462 students.

Data analysis

REGRESSION ANALYSIS EQUATIONS

The researcher best solves the prediction problem for future samples by *regression analysis equations,* the type of equations used by Suddarth and Wirt, Gorrell, and Breland and Jones. Simple regression analyses relate one predictor variable to one criterion variable (first-order correlation). Multiple-regression analyses are used if more than one predictor variable is used to predict a criterion variable (multiple-regression equations for multiple correlation). If combinations of multiple predictors are used to predict multiple criterion variables, canonical correlations are used. No other methods in the long history of psychology across a broad range of problems have ever been shown to be more accurate in their predictive capability than these types of statistical analyses (Meehl, 1954; Sawyer, 1966). It is important *not* to use expert judgment to set the prediction weights for the variables, because the predictions made by the regression procedures have never been exceeded and are cheaper to do.

The guiding principle is to introduce into the analysis *all* kinds of variables such as test scores, grades, or ratings of quality or success in classes. The best way to combine these variables in order to predict a criterion variable maximally is to use regression equations with regression weights, or B weights (including negative and zero weights), for each predictor variable. The analysis results in a predictor equation format. Suddarth and Wirt's equation using the raw score input as data was as follows:

6.2 (h.s. Rank)^3 + 4.3 (h.s. GPA) + 72.9 (SAT-V) − .22 (SAT-M) + 5.8 (Semesters of h.s. languages) − 8.9 (h.s. English grades reversed) + 8.5 = Purdue Freshman English grade

To calculate the *relative* predictive value of each predictor variable in the equation, a researcher essentially converts these variables to standard scores.

B WEIGHTS

B weights are those weights used when the variables are in raw-score or original form, in other words, have not been converted into standard score variables, which have means of zero and standard deviations of 1.00. Examples of such raw variables are typical test scores, grades, essay ratings, or GPAs. Each such variable gets an appropriate B weight in the equation according to its correlation and particular raw-score standard deviation. Because the raw scores generally do not have a mean of zero, the researcher must add a constant of adjustment to the set of B weights. This adjustment constant is labeled "A." Researchers must again be careful to use these B weights (and the A weight) only for prediction purposes and *not* for interpretation of the importance among variables in the equation. (Because all the variables with B weights are in raw-score form, their standard deviations often are quite different—that is, about .5 for GPA and about 100 for SAT scores. To weight these two variables equally the B weights would have to be 200 and 1, respectively.)

As was indicated above, for the variables that were in the original score form, Suddarth and Wirt calculated the B weights as follows: 6.2 for high-school rank cubed, 4.3 for high-school GPA, 72.9 for SAT-V, $-.22$ for SAT-M, 5.8 for semesters of high-school foreign language, and -8.9 for high-school English grades (reversed). They had to add a constant to the equation to adjust their several predictor scores to the mean of the criterion variable. In this case the constant, labeled A, was 8.5.

Note that Suddarth and Wirt's weights do *not* necessarily reflect the relative importance of each of the variables in the prediction because, as said above, these predictor variables generally have widely different standard deviations—for example, 100 for the SAT-V and SAT-M, probably .4 to .5 for high-school GPA. The B weights for each variable have accommodated for differences in the standard deviations of the predictor and criterion variables. To interpret the actual weights for comparison of the relative importance of each variable in the prediction, a researcher needs *beta weights*.

BETA WEIGHTS

Beta weights are the weights used for predictor and criterion variables when they are stated in standard score form, for example, all with means set at zero and standard deviations set at 1.00. Beta weights should be used rather than B weights to determine the *relative* importance of the several predictor variables in the multiple correlation. However, even when using beta weights for the predictor variables, several factors can change the beta weights markedly in prediction equations: addition or deletion of predictor variables, changes in their reliabilities, preselection of the subject sample on a variable, and so forth. Breland and Jones' final equation with the beta weights for all significant predictors was as follows:

.37 (ECT Objective score) + .28 (Essay length) + .12 (Overall organization) + .10 (Paragraphing and transitions) + .07 (Noteworthy ideas) + .05 (Parallel structure) + .08 (Spelling) + .05 (Precision of diction) = ECT holistic score

Because all the variables have the same standard deviations, the beta weights can be compared among themselves for relative strength of prediction in the equation: the bigger the beta weights, the more important the variable is for a particular prediction. However, because of the intricate interrelationships among the predictor variables in the equation, it is difficult to use beta weights to determine the *theoretical* importance of individual variables in relation to the criterion. In other words, researchers cannot interpret a beta weight for even one of the predictor variables in isolation from the rest. Prediction equations generally are useful only to make specific predictions or placements, not for use in interpreting the theoretical relationships among variables.

A computer calculates the prediction equations and automatically prints the correlation matrix. As variables are added to, or subtracted from, the equation, the computer also prints successive multiple correlations, the B weights and the constant A, and the beta weights for each entered predictor variable. (For the analysis techniques, see any intermediate-level statistics book; Nie et al., 1979; Cooley and Lohnes, 1971, p. 195; Guilford and Fruchter, 1973; Hopkins and Glass, 1978.)

In Suddarth and Wirt's study, we know that the SAT-V had a major weight in the prediction equation because of its known large standard deviation, in relation to other variables, of near 100 and its weight of 72.9. (This standard deviation would actually be somewhat reduced from the general population value because Purdue restricts the range of its students to the upper levels of the SAT-V, SAT-M, and high-school GPA.) Alternatively, the high-school GPA standard deviation in raw-score form was probably about .4 and yet was still only weighted 4.3. Suddarth and Wirt calculated the beta regression weights for each predictor variable in the matrix. All four mathematical and science high-school grades essentially weighted zero and thus were dropped from the equation, therefore simplifying it.

Gorrell used a standard multiple-correlation SPSS program to analyze her data. The regression analysis was stepwise, indicating that she entered the initially highest correlating predictor variable with the criterion variable into the regression equation; then, after adjusting the correlation matrix for the effect of entering the first variable, she entered the next highest of the residual correlations, and so on. (The process can be stopped at any point.) Further, the multiple correlation was squared and this value was used to indicate the percentage of variance (the individual differences) explained on the criterion available. All these calculations and entries were printed automatically as each new variable was entered into the multiple-regression equation via the standard computer program.

Gorrell ultimately developed a relatively simple regression equation. She pointed out, rightly, that after using the Pre-essay scores and the ACT English scores, she gained little by adding all the other seven measures she had collected. Further, by using the regression equation in reverse and inserting a criterion grade of "C," she estimated the Pre-essay grades and ACT English scores that would be required to attain a "C." (It would have been helpful if she had presented the actual regression equation.) She also found differences in the prediction measures used between a standard freshman composition and a basic writing skills course.

Note that Gorrell did *not* try to explain why certain variables did not enter the regression equations in the stepwise analysis. As was indicated, regression analysis should be used solely for prediction purposes and not for theoretical interpretation or description. On the basis of her analyses, she also made a number of recommendations to her department regarding grading standards, student allocation, and so forth, which were adopted and used experimentally.

Gorrell is to be commended on the completeness of her statistical presentation. She gave the complete correlation matrix and the multiple-correlation entry table (see Tables 6-1 and 6-2.) But there was a problem: her ratio of subjects to predictor variables was small. She indicated that 96 students were in one group of classes and 48 in the basic writing course. With the ten original variables, the number of students was too small to achieve relatively stable results in the basic writing course (see Tables 6-1 and 6-2).

SHRINKAGE

The major assumption in prediction research used for placement is that a future sample of people (and judges for ratings and grades) will be similar to the ones on which the prediction equations were used. It is wise periodically to repeat the research that developed the regression equations in order to determine if any shifts of judgment have occurred or if the samples of students have changed, both of which can change

Table 6-1 Multiple-Regression Summary with Post Essay Rating as Criterion Variable

Variables	R-Square	RS Change	Beta	F-Value (beta)
Pre Essay Score	.422	.422	.346	7.164*
ACT English	.484	.062	.424	10.199*
ACT Social Studies	.501	.017	−.156	2.029
Essay Errors	.508	.007	.095	.882
Reading Efficiency	.511	.003	−.056	.447
"Aluminum" Errors	.512	.001	−.039	.158
Cloze Errors	.513	.000	−.026	.049
Reading Comprehension	.513	.000	−.040	.085
Reading Vocabulary	.513	.000	.031	.057

Source: Gorrell (1983), Table 1. Copyright © 1983 by the National Council of Teachers of English. Reprinted by permission of the publisher.

n = 103
*Statistically significant: $p < .01$.
R = multiple correlation coefficient
R-Square = the multiple correlation coefficient squared
R S Change = the change between the prior R Square and this one
Beta = the weight of this variable when it is in standard score form
F-Value = the statistical test for significance (from zero) of beta

Table 6-2 Correlation Coefficients of Selected Predictor Variables and Post-Holistic Rating of Essays

Variables	ACT-E	ACT-S	Rdg-V	Alum	Cloze	EsErr	HolR1	HolR2
ACT English	1.000	.620	.656	−.526	−.738	−.489	.649	.611
ACT Social Studies		1.000	.608	−.177	−.519	−.275	.348	.254
Reading-Vocabulary			1.000	−.379	−.598	−.395	.656	.461
"Aluminum"				1.000	.542	.543	−.527	−.467
Cloze					1.000	.477	−.650	−.527
Essay Errors						1.000	−.640	−.505
Holistic Rating (Pre)							1.000	.649
Holistic Rating (Post)								1.000

Source: Gorrell (1983), Table 2. Copyright © 1983 by the National Council of Teachers of English. Reprinted by permission of the publisher.

n = 103
$p < .05$ when $R > .195$.
$p < .01$ when $R > .254$.

R = multiple correlation coefficient
R-Square = the multiple correlation coefficient squared
R S Change = the change between the prior R Square and this one
Beta = the weight of this variable when it is in standard score form
F-Value = the statistical test for significance (from zero) of beta

the weighting of the predictor variables. If the samples and measurements remain relatively stable, further research will tend to yield more stable weights over the years.

It is well known that multiple-prediction equations must be "preshrunk." Researchers develop prediction equations on sets of data for samples that already have passed through a course or an educational program. When researchers shift from the prediction-establishing sample to future samples, the equation usually does not predict quite as well—especially if the ratio of the original number of subjects to the number of variables was not large. The overfitting of the data that results is reduced in the future samples, and the multiple correlation as a result is also reduced—that is, "shrunk." (Equations exist to approximate what this shrinkage will be; see Guilford and Fruchter, 1973.)

To see how well the 1971 predictions would have done on the new sample in a real prediction situation, Suddarth and Wirt used the 1971 equation on data from freshmen entering in 1972. This constituted a way of validating their prediction equation. On the 1972 data for the overall combined and adjusted English course grades, the multiple correlation was only .62 as compared with the 1971 correlation of .79. This shrinkage was expected. As long as the samples are fairly large and the ratio of ten subjects to each variable holds, the shrinkage will not be great. It is important to redo prediction studies periodically because of changes in the nature of student bodies, curricula, and grading standards.

Research design: classification

When a researcher is trying to predict a nominal criterion variable like academic major instead of an interval criterion variable like grade, the forecasting research is considered a classification study. Researchers can determine how well an individual will fit into certain categories like business major, English major, or engineering major. They use *discriminant analysis* to predict an individual's compatibility with one of these categories. For a nominal criterion variable like preference for major, the researcher designates several possibilities (e.g., English, engineering, political science) and classifies each student as an "0" or a "1" on each criterion variable in the set. The resulting matrix allows the researcher to predict the probability of the subject's being in each of the categories. The computer conducts these discriminant analyses. (See Nie et al., 1979; Cooley and Lohnes, 1971.)

In classification studies, the results are similar to those of prediction.

The researcher develops regression weights for the current sample of subjects indicating which of the predictor variables maximally predict each major. Then the researcher generates "hit-and-miss" tables to show which categories each person "looks most like" as a result of these sets of equations.

This research should be replicated on several samples to be relatively definitive. Because a researcher develops a regression equation for each category in the nominal criterion variable, the number of subjects should be ten times the number of predictor variables plus all the categories of the criterion variable in order to yield stable results.

Cautions for prediction and classification research

Problems with subjects

As indicated above, one precaution a researcher or reader must take is to be wary of the n to k ratio, the ratio of the number of people in the sample to the number of variables to be used as predictors or criteria. This ratio should be at least ten subjects to one variable (although it can be reduced to as low as 5:1 as the population approaches several thousand). If the ratio is smaller, the multiple correlations will be too large as a result of an overfit of the regression equations. These correlations will therefore tend to shrink in future samples, causing later use of the old regression equation with new samples to yield predictions smaller than those found in the original samples.

Another problem is restricted ranges of variables in samples, which can cause researchers to misinterpret a correlation based on such a sample. For example, faculty in selective colleges may note that correlations are only around $+.25$ between examinations like the SAT predicting success in college and grades in freshman English. This relationship predicts only 6.25% ($.25^2 \times 100 = 6.25\%$) of the variances (individual differences) of these grades. Teachers in such select schools may then exclaim that 93.75% of the variance of the grades is not being predicted and therefore recommend not using the tests to admit students. Little do they realize that their initial selection of students on the basis of the SAT in itself reduces the observed correlations of the SAT and grades of those who are actually in the college receiving grades. Therefore, care must be taken in interpreting correlations based on samples that restrict the range of any of the variables being correlated.

Problems with measurement

In many cases a researcher has only one criterion variable such as a grade, or a few related variables, such as major paper and final examination grade. In multiple-criteria cases, however, the researcher should examine the relationships among these criteria. If the correlations are positive, then the researcher can predict either a main criterion variable or an average of the criterion variables. As an alternative approach, the researcher can predict each criterion variable separately and then examine each predictor weight across all criteria to ascertain similarities across all the predictor equations. If the weights are similar, then perhaps an average of the separate weights by predictor variables is appropriate. (If researchers find that criterion variables that reflect course success are relatively independent, then the goals of the course should be examined for cohesiveness.)

Care should also be taken not to add criterion variables together naively to form an overall criterion. Interval variables in the behavioral sciences usually are not additive in a simple, equally weighted fashion. The general rule in adding variables together is that when variables are combined, they are self-weighting as functions of their own variances. The larger the individual differences in a variable, the more it weighs in a combination of variables. While this caution about adding variables may seem unrelated to the concerns of most English instructors, they do in fact engage in a form of such combining when they add grades for tests, papers, and class participation to obtain an overall course grade, inadvertently weighting the papers or tests with the greatest individual differences the heaviest.

Further, researchers should use the original data variables instead of combining them by using subtraction, ratios, or quotients. Simple subtraction, ratios, and quotients of variables assume that the correlation between these variables is 1.00; this can never be the case with less than perfect reliabilities, which is the case for all behavioral science variables. If a researcher then goes on further to correlate variables that are the result of subtractions, ratios, and quotients, with other variables, strange results will often occur (Asher, 1976).

INFLUENCES ON CORRELATIONS

Before combining variables via their correlations for prediction purposes, a researcher would do well to look for possible influences on the correlation coefficients. Recall that correlations show the strength of relationships between variables in a sample of people. However, the apparent size of a relationship in a sample can vary in ways different from the true relationship between these same variables in a general population. Several factors influence the correlation.

1. Curvilinear-based correlations are higher than the usual linear correlations that are calculated.
2. The internal-consistency reliability of the measurement affects the possible correlation between variables: the higher the reliability, the higher the correlation (see Appendix I).
3. The use of dichotomized variables tends to reduce correlations with these variables because they tend to be less reliably measured than multiple-category variables.
4. A split in dichotomous variables that is other than 50% does not yield as much correlation.
5. The correlations of averages of variables over groups of people or classes tend to be much higher than the correlations of the individuals' scores over the variables (aggregate correlations).
6. The use of percentile scores or rank orders (assumed to be equal intervals) distorts the correlations among variables from what they would be if true interval measurement quantification systems were used.
7. Restrictions in the range of a variable because of a selective population reduce the correlation between that variable and all other variables.

Problems with interpretation

A final caution is in order. When two or more predictor variables are used to predict success in a college course or some other criterion variable, the researcher is sometimes tempted to interpret the resulting weights on each predictor value as true weights that help establish a descriptive theory capable of interrelating all the variables. For example, a researcher may erroneously assume that SAT scores in isolation are the best predictors of success in a composition course because they have the greatest weight in a prediction equation. This is hazardous, because the sole purpose of multiple correlation is to predict maximally a criterion variable. Because the analysis automatically adjusts the weights of each predictor variable in relationship to one another as well as the criterion variable, an attempt to interpret the overall prediction equation weights among the variables for purposes of establishing theory will often be in error. (This is a common problem in the research literature.)

The simple first-order correlations among variables will vary some-what from sample to sample, variations that may markedly change the order of entry and the various weightings of the predictor variables in the equation. Thus a researcher needs to examine the multiple-correlation weightings of the variable across a series of samples to get a better idea of the weights for a given variable as a predictor of a criterion variable.

Summary

The purpose of prediction is to determine the relationship between a set of predictor variables and the interval criterion variable; classification studies show similar interrelationships and predict status on a nominal criterion variable. The researcher introduces and correlates all types of data as predictor variables and criterion variables on the subjects (usually at least ten times the number of variables), by means of regression analysis or discriminant analysis in order to determine which variables will best constitute a formula predicting a subject's future scores or status. Care, however, must be taken to understand the influences on these correlations.

Such prediction and classification studies are highly valuable for institutions that wish to place students in appropriate courses or programs, or for researchers who wish to determine the set of predictor variables that will account for a specific criterion variable. In fact, these principles are so well established in prediction research methods that they call into question the ethics and legality of placing or classifying students or predicting their success in composition classes in any other way.

Checklist for reading "Prediction and Classification Studies"

1. Who were the subjects? How were they selected?
2. For what purpose was the study conducted?
3. What were the predictor variables?
4. What were the criterion variables?
5. What analyses were used?
6. What was the prediction equation?
7. What, if any, validation studies were conducted?
8. What, if any, problems affected the study?

References

Research Design and References

American College Testing Program (1975). *Your college freshman*, 5th edition. Iowa City: The American College Testing Program.

Asher, J.W. (1976). *Educational research and evaluation methods*. Boston: Little, Brown.

Cooley, W.W., & Lohnes, P.R. (1971). *Multivariate data analysis*. New York: Wiley.

Cramer, S., & Stevic, R. (1971). A review of the 1970–71 literature: Research on the transition from high school to college. *College Board Review, 81,* 32–38.

Duggan, J. (1963). Field testing a central prediction service. *College Board Review, 49,* 12–15.

Epstein, S. (1979). The stability of behavior: I. On predicting most of the people much of the time. *Personality and Social Psychology, 37,* 1097–1126.

Fishman, J. (1958). Unsolved criterion problems in the selection of college students. *Harvard Educational Review, 28,* 340–349.

Guilford, J.P., & Fruchter, B. (1973). *Fundamental statistics in psychology and education*. New York: McGraw-Hill.

Hopkins, K.D., & Glass, G.V (1978). *Basic statistics for the behavioral sciences*. Englewood Cliffs, N.J.: Prentice-Hall.

Jones, M.R. (Ed.) (1970). *Miami symposium on the prediction of behavior, 1968*. Coral Gables: Univ. of Florida Press.

Meehl, P.E. (1954). *Clinical versus statistical prediction*. Minneapolis: Univ. of Minnesota Press.

Munday, L.A. (1970). Factors influencing the predictability of college grades. *American Educational Research Journal, 64,* 148–156.

Nie, N.H., Hull, C.H., Steinbrenner, K., & Bent, D.H. (1979). *SPSS: Statistical package for the social sciences*, 2nd edition. New York: McGraw-Hill.

Sawyer, J. (1966). Measurement and prediction, clinical and statistical. *Psychological Bulletin, 66*(3), 178–200.

Taylor, H.C., & Russell, J.T. (1939). The relationship of validity coefficients to the practical effectiveness of tests in selection: Discussion and tables. *Journal of Applied Psychology, 23,* 365–378.

Prediction and Classification Studies

Breland, H.M., & Jones, R.J. (1984). Perceptions of writing skills. *Written Communication, 1,* 101–119.

Chase, D.D. (1984). An English placement examination: An analysis of the effectiveness of the examination as used at five junior colleges in California. *DAI, 45,* 02A.

College Entrance Examination Board (1975). *The test of standard written English: A preliminary report*. ERIC Document Reproduction Service No. ED 117162.

Creedden, J.E. (1984). The effect of high school writing experiences on scores on the University of Wisconsin English placement test. *DAI, 45,* 12A.

Daly, J. (1978). Writing apprehension and writing competency. *Journal of Educational Research, 72,* 10–14.

Daly, J., & Miller, M. (1975). Further studies on writing apprehension: SAT scores, success expectations, willingness to take advanced courses and sex differences *RTE, 9,* 250–256.

Godshalk, F.I. et al. (1966). *The measurement of writing ability*. New York: College Entrance Examination Board.

Gorrell, D. (1983). Toward determining a minimal competency entrance examination for freshman composition. *RTE, 17,* 263–274.

Grobe, C. (1981). Syntactic maturity, mechanics, and vocabulary as predictors of quality ratings. *RTE, 15,* 75–85.

Harris, M. & Wachs, M. (1986). Simultaneous and successive cognitive processing and writing skills: Relationships between proficiencies. *Written Communication, 3,* 449–470.

Hartman, D.M. (1984). An investigation into the predictive relationship of ten writing assessment variables to reading comprehension. *DAI, 45,* 05A.

Hoffman, R.A., & Smith, D.L. (1979). The use of "Test of Everyday Writing Skills" in a college screening program. *Journal of Educational Research, 73,* 168–171.

Illo, J. (1976). From senior to freshman: A study of performance in English composition in high school and college. *RTE, 10,* 127–136.

Johnson, N. (1977). The use of grammatical and rhetorical norms, pedagogical strategies, and statistical methods in designing and validating a composition placement instrument. *DAI, 37,* 11A.

Lamb, B. (1984). The prediction of freshman composition grades at a community college: A correlational study based on a non-computational readability scale. *DAI, 45,* 04A.

Lenning, O.T. (1975). *Predictive validity of the ACT tests at selective colleges.* Research Report No. 69. Iowa City, Iowa: The American Testing Program.

Mosenthal, P., & Na, T.J. (1981). Classroom competence and children's individual differences in writing. *Journal of Educational Psychology, 73,* 106–121.

Mosenthal, P., Davidson-Mosenthal, R., and Kreiger, V. (1981). How fourth graders develop points of view in classroom writing. *RTE, 15,* 197–214.

Nold, E., & Freedman, S. (1977). An analysis of readers' responses to essays. *RTE, 11,* 164–174.

O'Keefe, B., & Delia, J.G. (1979). Construct comprehensiveness and cognitive complexity as predictors of the number and strategic adaptation of arguments and appeals in a persuasive message. *Communication Monographs, 46,* 231–240.

Page, E. (1968). The use of the computer in analyzing student essays. *International Review of Education, 14,* 253–263.

Passons, W. (1967). Predictive validities of the ACT, SAT, and high school grades for first semester GPA and freshman courses. *Educational and Psychological Measurement, 27,* 1143–1144.

Pelligrini, A. (1984). Symbolic functioning and children's early writing: The relations between kindergartners' play and isolated word-writing fluency. In R. Beach and L. Bridwell (Eds.). *New directions in composition research.* New York: Guilford Press, pp. 274–284.

Rowan, K.E. (1985). *Producing written explanations of scientific concepts for lay readers: Theory and a study of individual differences among collegiate writers.* Doctoral dissertation, Purdue University.

Rubin, D., Piché, G., Michlin, M., & Johnson, F. (1984). Social cognitive ability as a predictor of the quality of fourth-graders' written narratives. In R. Beach and L. Bridwell (Eds.) *New directions in composition research.* New York: Guilford Press, pp. 297–307.

Sawyer, R., & Maxey, J. (1975). *The validity over time of college freshman grade prediction equations.* Research Report No. 80. Iowa City, Iowa: The American Testing Program.

Singer, M.M. (1984). An integrative model of competent writing. *DAI, 45,* 08A.

Stewart, M., & Grobe, C. (1979). Syntactic maturity, mechanics of writing, and teachers' quality ratings. *RTE, 13,* 207–215.

Stewart, M., & Leaman, H. (1983). Teachers' writing assessments across the high school curriculum. *RTE, 17,* 113–125.

Suddarth, B., & Wirt, S.E. (1974). Predicting college placement using precollege information. *College and University, 49,* 186–194.

Washington, S.N. (1979). A study to ascertain the relationship between pre and post Nelson–Denny reading test scores of 275 freshmen and their reading and writing grades at a suburban state college. *DAI, 39,* 4708A.

Weiss, K. (1970). A multi-factor admissions predictive system. *College and University, 46,* 203–210.

7

Measurement

The process of quantifying variables is called measurement. In this chapter, we present the principles governing measurement. A number of collections of essays recently have dealt with writing assessment or evaluation of student texts, especially for grading or placement purposes (e.g., Cooper and Odell, 1977; Cooper, 1981; Stiggins, 1982; Spandel and Stiggins, 1981; White, 1985; Greenberg, Weiner, and Donovan, 1986). This chapter's focus is on writing evaluation as a measurement tool for research.

Writing researchers have essentially two measurement options. They can select currently available measures that often have normative standards, reasonably known reliabilities, and some evidence of validity. It is always wise to use available standard instruments, tests, attitude scales, observation coding schemes, and questionnaire items if they are reliable and valid for the researchers' purposes. Researchers gain a number of advantages by using these available instruments. Someone else has already established their validity and reliability. Researchers can also compare their results with those of other studies using these instruments. These measures, therefore, save time and money and lend credibility to an individual's research.

Sometimes, however, there are no existing instruments to assess validly the aspects of writing that an investigator wants to measure. In these cases, the researcher needs to develop new instruments—either test items, writing tasks to be graded, or observational rating scales of various kinds. But developing a good measurement instrument is enormously time-consuming and often requires as large an effort or greater than that needed to plan and execute the research design. For example, training expert judges to rate series of compositions or other writing variables using a new instrument is an expensive and lengthy process. But the available standard measures may be too gross (Quellmalz, 1981; Scardamalia and Bereiter, 1985) to measure certain composition variables of interest like planning behaviors or audience adaptation, for which a researcher may have to devise new instruments.

Researchers must also decide whether to use direct or indirect measures of writing. Stiggins (1982) discusses the merits and limitations of both types of measurements. Some of the frequently used direct instruments include holistic, analytic, and primary trait scoring of essays and T-unit analysis. Indirect measures include various standardized tests of grammar, punctuation, style, or usage. In Appendix II, we provide a list of measurement instruments that can be used in research on writing.

Composition measurement instruments

To help explain the principles of measurement, we briefly characterize here several frequently used instruments: holistic, analytic, primary trait scoring, standardized tests, T-unit analysis, and attitude scales. For *holistic evaluation,* the rater assigns a single rank or score to a piece of writing, either grouping it with other graded pieces or scoring it on the basis of a set scale. The holistic evaluator, having been trained with other raters, gives a rank or score impressionistically and does not enumerate features of the piece of writing. The English Composition Test of the College Entrance Examination Board (CEEB) is an example of holistic evaluation (Kirrie, 1979).

For *analytic evaluation,* the rater gives a separate score for each of several qualities of a piece of writing. The Diederich Analytic Scale, for example, tests the following features: general merit (ideas, organization, wording, and flavor) and mechanics (usage, punctuation, spelling, and handwriting) (Diederich, 1974).

For *primary trait scoring,* the rater gives scores based on criteria that have been established by analyzing the assignment and the resources used by writers in fulfilling it. In other words, the criteria, unlike holistic and analytic scoring, vary from one type of writing to another and hence are rhetorically based. Primary trait scoring is used by the National Assessment of Educational Progress (Lloyd-Jones, 1977, 1987).

Standardized tests, often multiple choice, are objectively scored tests that are usually machine-scored. They often assess knowledge of usage, grammar, punctuation, and syntax. A widely used example is the Test of Standard Written English of the College Entrance Examination Board (CEEB, 1975).

T-unit analysis involves counting the number of T-units (one main clause plus any subordinate clauses attached to it) in a piece of writing and computing the average length of the T-unit by counting the number of words in forty-five T-units selected from three or more samples of a given student (Hunt, 1965; Mellon, 1969; Cooper, 1975).

Attitude scales often use the Semantic Differential (Osgood, Suci, and Tannenbaum, 1969) or a Likert-type scale on a series of questions about writing or its pedagogy. The Schuessler, Gere, and Abbott Teacher Attitude Scale and the Writing Apprehension Test are examples of this type of measurement (Schuessler, Gere, and Abbott, 1981; Daly and Miller, 1975a and 1975b).

To assess *writers' knowledge*, new instruments are being developed. Faigley, Cherry, Jolliffe, and Skinner (1985) discuss Performantive Assessment instruments they have developed to measure changes in both writers' composing strategies and their knowledge in such tasks as classification, inductive argument, deductive argument, hypotheses construction, and audience adaptation. They also review other knowledge-assessment instruments. Langer (1984) and Newell and Mac-Adam (1987) have devised Topic-Specific Knowledge measures. Murray (1987) has developed two measures to assess the extent to which students reach new insight through writing expressive discourse. Connor and Lauer (1985; 1988) have developed a Persuasive Appeals Analysis Scale to measure the levels of rational, credibility, and affective appeals in persuasive texts and a Toulmin Analysis of Informal Reasoning.

A number of articles and books provide good overviews of measurement instruments for writing (Cooper and Odell, 1977; Cooper, 1981; Odell and Cooper, 1980; Stiggins, 1982; Fagan, Cooper, and Jensen, 1975 and 1985, and Breland et al. 1987).

Types of measurement

In science, there are four levels of measurement: nominal, ordinal, interval, and ratio. In most of the behavioral, cognitive, and social sciences, however, only three types of variables are widely used—nominal, ordinal, and interval. *Nominal* measurement, sometimes not considered measurement at all, is simply counting frequencies of occurrences of behavior or conditions in various classifications such as types of writing: essays, poems, novels, biographies, and so forth. It is a type of measurement often used in qualitative descriptive research, as we have seen. In determining if a piece of writing fits in one of the classifications, a researcher simply indicates yes or no. If quantification is necessary, the piece is numbered 1 or 0.

Levels

ORDINAL

An *ordinal* scale places people or themes in an ordered relationship to one another on one dimension or variable at a time. If the dimension is quality of writing at the end of a course, then raters rank papers not by distance or interval between students, but in some order of quality: the best paper, the next best, and so on. Writers also can be ordered as remedial, regular, or advanced. This rank ordering does not necessarily consider the writing or the students to be equally distant from one another.

INTERVAL

Interval measurement has all the properties of rank ordering but in addition the interval levels are expected to be equally distant from one another. Many ratings, judgments, grades, and test scores in composition can be considered interval measurement—if the operational definitions used in developing them are based on intervals that are equal-appearing to expert or experienced judges. Most grade scales (A–F) and holistic scales (1–5), if developed with care, are equal-interval scales. The Semantic Differential uses a seven-point scale to measure attitude toward polar terms (Osgood, Suci, and Tannenbaum, 1969). The Likert Scale uses a five-point scale. Interval measures are important to achieve because many of the most powerful and frequently used statistical analyses assume equal-interval quantification systems.

RATIO

Ratio measurements have all the properties of interval measurement plus a true zero, a fixed point at which none of the dimensions exists. This true-zero point exists in the physical world as absolute zero in temperature, no weight, or no distance, but it is difficult, perhaps impossible, to define for a purely psychological variable. In most statistical analyses, however, ratio scales are unnecessary. Composition research does not use ratio scales for the most part.

Scores

STANDARD SCORES: Z SCORES

Interval measurement systems must be developed carefully to have some assurance that the resulting scores reasonably meet the assumption of equal intervals. Most standardized tests meet the goal of producing equal scores because they have been developed so that when groups of people (like those on whom the test was normed) take the test, their scores will be considered equal-interval data. (See Appendix I for more information about standard scores.)

Standard scores result from taking the raw scores on a test, subtracting the mean of the scores, and dividing by their standard deviation. This results in a set of scores that have a mean of zero and a standard deviation of 1.00. These basic standard scores are called "z scores" and are the basis of several other types of standard scores that are widely used in the behavioral sciences and education. The z score is very useful also in comparing the results of experimental research. For instance, the difference between the mean of an experimental group and the mean of a control group can be divided by their averaged standard deviations and expressed as a z score. This is exactly what is done in meta-analysis (see Chapter 10). Standard tests like the Scholastic Aptitude Tests multiply the basic z scores by 100 and add 500 to all. This gives a mean of 500 and a standard deviation of 100, an advantage over z scores of no negative scores and no need for decimals.

PERCENTILE SCORES

Percentile scores, while rank ordered, are *not* equal-interval scores when compared to any of the standard score metrics (see Appendix I). This distinction is not too much of a problem for scores not too far from the mean of the distribution. It does become of concern if one is working with gifted, talented, or remedial populations. The percentile scores that appear close together at the upper and lower ends of the population are actually rather enormous differences psychologically. Hence, it is better not to use rank ordering or percentile score systems for data analysis.

Many ratings, grades, interviews, and questionnaire categories can yield interval measurements if the ratings or categories are well constructed. The responses must be related to only one dimension of behavior, and the response categories to questions should be ordered in some way that is theoretically meaningful. If these successive, ordered categories are developed in such a way that the differences between adjacent categories are equal-appearing intervals, then interval numbers can be applied to the categories and more powerful and higher order statistics can apply. For example, the standard phrases of the Likert Scale—"strongly agree," "agree," "uncertain," "disagree," and "strongly disagree"—have been used to describe a successive-category response system. These phrases have been shown to constitute equal-appearing intervals in representative groups of responders. Thus, an equal-interval numbering system can be appropriately applied to these specific phrases (Likert, 1932).

Reliability

Reliability, one of the main requirements of all measurement instruments, is fundamental to the process of quantifying variables in the behavioral and social sciences. Essentially it is the ability of independent observers or measurements to agree. Without agreement in the observation of a subject, a group, a criterion, or a treatment, a researcher can make little progress in research. Thus reliability is to a large degree a social construction, a collaborative interpretation of data, influenced, as Polanyi (1958) has shown, by researchers' tacit knowledge. Agreement among independent observers helps a researcher to generalize results and make claims about replicability.

The reliability of a measurement is generally reported as a decimal fraction, stated in terms of a positive correlation coefficient that will range from .00 to close to 1.00. Correlation, a statistic usually abbreviated as r, can take the values of -1.00 to 0 to $+1.00$. An r value of $+1.00$ means there is a perfect straight-line (or linear) agreement and perfectly precise agreement between two sets of observations; an r of .00 means there is no linear relationship between sets of observations; an

r value of -1.00 means that there would be absolutely reversed agreement among observers—high scores with low scores and low scores with high scores.

Types of reliability

There are three major types of reliability: equivalency, stability, and internal consistency.

EQUIVALENCY

Equivalency reliability is usually determined by relating two sets of test scores to one another (in that the problem is to show that one test is essentially equivalent to another for a given group of subjects). The reliability is simply the degree of relationship between the two sets of scores for a given sample of people who take both tests at the same time. The averages, standard deviations, and average intercorrelations among items must be the same as well. The scores of people on Test A would simply be correlated with their scores on Test B. In scatter-diagram form, these individuals might plot as in Figure 7-1 to show that the total scores of the two tests are highly related. Equivalency reliability is reported in terms of a reliability correlation coefficient.

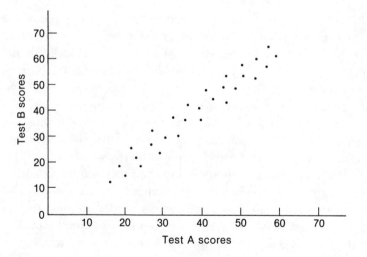

Figure 7-1 Plot of Test A and Test B scores for twenty-eight subjects with an equivalency reliability of about $r = .85$.

STABILITY

The stability of measurement instruments rests on their agreement over time. A holistic, analytic, or primary trait analysis of students' writing would be stable, if at different periods (when there is no good reason for changes) raters scored writing performances to be relatively the same.

To determine the stability of a measure, researchers repeat the same tests (or observations) on the same subjects at some future date, say a month, six months, or a year or more after the first test. They then correlate the test scores from one time with the test scores from the other time for the same subjects (much as the scores from Test A and Test B were correlated in Figure 7-1). The resulting correlation is the measure of stability for the given period of time, the given sample of subjects, and the measurement procedures simultaneously. Witte and Davis (1980, 1982), for example, studied the stability of mean T-unit length within and across different types of discourse (description, narration, classification, and comparison) with individuals and groups. In their second study they found the T-unit to be a stable individual trait as well as a stable group characteristic. Daly and Miller (1975a) also tested for stability by administering retests to assess the stability reliability of the Writing Apprehension Test. They found it to have a reliability of $r = .92$.

INTERNAL CONSISTENCY

The theoretically most important type of reliability is internal consistency, which is essentially an index of the precision of the measurement instrument or of the observers. This kind of reliability helps the researcher interpret data because it makes use of and is the basis of the theory of measurement, which predicts the expected values of scores and the limits on the degree of relationship among variables. It helps researchers understand why all observed scores (except those at the average) are biased, why all observed correlations among variables are underestimates of the true correlations, and why all observed standard deviations are too large as compared to their true score relationships.

If test items or observers are in agreement upon assessing a single dimension of human ability or achievement using ratings, the test or observers are said to be internally consistent. For example, in the case of a test of writing mechanics that uses true–false or multiple-choice responses, each item is considered as a miniature test in itself. All possible correlations among all pairs of these test items can be made and, in a sense, averaged to indicate how precisely the items measure a single dimension or aspect of writing mechanics. If the correlations among all the possible pairs of items tend to be positive and about as large as they can be expected to be, then the items can be said to be measuring a single dimension of human achievement. Such a test is then said to have high internal-consistency reliability. Kuder and Richardson (1937) developed an equation to assess the internal-consistency reliability of dichotomously scored items only. Their effort resulted in a well-known internal-consistency reliability equation called the KR20. Internal-consistency reliability is usually assessed now by means of Cronbach's alpha, which is appropriate for both interval data and dichotomously scored data.

In order to develop scales of teacher attitudes, Schuessler, Gere, and Abbott (1981) applied Cronbach's alpha to 46 items in five scales that they took from a Composition Questionnaire developed by the Commission on Composition of the National Council of Teachers of English. Because nine items of one scale had a low reliability, they eliminated this scale and used four scales with the following alpha coefficients: (1) attitudes toward the instruction of the conventions of standard written English, $r = .72$; (2) attitudes toward the development of students' linguistic maturity, $r = .73$; (3) attitudes toward defining and evaluating writing tasks, $r = .70$; and (4) attitudes toward the importance of student self-expression, $r = .74$. Daly and Miller (1975) found a Cronbach's alpha coefficient of .94 for the Writing Apprehension Test.

Another way of computing the internal consistency of a test is to correlate the scores of two halves of a test (odds versus evens) with each other. This is called split-half reliability. The Educational Testing Service, for example, determined the internal-consistency reliability of the Test of Standard Written English to be .87 (CEEB, 1975), based on the correlation between items in one half of the test with items in the other half of the test. Daly and Miller (1975a) also computed the split-half, internal-consistency reliability of the Writing Apprehension Test, finding a coefficient of .94. Computer programs can be used to print split-half internal-consistency reliabilities using correlations between the total of the odd items and the total of the even items across all subjects.

If researchers had to calculate internal-consistency reliabilities in a test of more than several items, it would obviously be a lengthy process to determine all possible correlations between pairs of items or raters. Today, however, the computer does it quickly (though with a somewhat different approach). Most equipment for scoring tests used in schools and universities will automatically yield internal-consistency correlations in addition to each person's raw and standard scores, frequency distributions of scores, class means, and standard deviations.

Determining internal consistency also helps researchers to check the viability of an analytic scale, in which each quality or category is treated as a separate writing ability such as organization or ideas. Because a variable is a conceptual entity, it should represent only one aspect of a phenomenon of interest. If all categories of an analytic scale have a high correlation with one another, they really do not represent or measure different abilities. With an overall high internal consistency, they may measure only one variable—perhaps a general writing ability.

Interrater Reliability

In composition studies, a common form of internal consistency is interrater reliability—the amount of correlation between two or more raters or coders. To determine the internal-consistency reliability for intervally scored items such as analytic ratings of ideas, organization, or interval holistic ratings of overall quality as expressed by several judges, researchers can use the Cronbach's alpha internal-consistency reliability in such packages as the Statistical Package for Social Sciences (SPSS).

To achieve high interrater reliability, raters must be carefully trained, using similar data other than that to be rated in the research. Good operational definitions are essential in helping raters to reach agreement. When they have achieved a high level of agreement in training, they can then begin assessing the research data. A researcher can report the reliability coefficient for agreement in two ways: either as a coefficient for a percentage of randomly selected pieces of data or for the entire set of data.

If the data are not in interval form, or if simple rater agreements are desired over a series of checklist items (such as "use of period is correct"), then a simple percentage of rater agreement may be used. (To be interpreted adequately, this percentage agreement of raters should be compared to the expected percentage agreement by chance alone. This agreement is not as useful theoretically as the internal-consistency reliability, but it does indicate that more than one observer agrees that the perceived phenomenon does exist.) To correct for chance agreement, a researcher can use Cohen's kappa (see Appendix I).

Important also is the way in which ratings or codings are used to obtain a score. The best way is to *average* the ratings in order to report as accurate a result as possible. Forced consensus does not yield as accurate a score, because some dimensions of disagreement are thereby obscured.

Currently opinions differ as to the level of reliability needed for different purposes. Cooper, citing Diederich, maintained that "a reliability coefficient of .80 is considered high enough for program evaluation, a reliability coefficient of .90 for individual growth measurement in teaching and research" (Cooper, 1977, p. 18). He cited some studies that have achieved interrater reliabilities of from .81 to .95 for five raters on five different types of holistic ratings (Follman and Anderson, 1967) and .95 for three raters of "creative writing" (Moslemi, 1975). Finally, he cautioned that using only one writing sample does not yield high reliabilities (Cooper, 1977, pp. 18–22).

Nunnally (1978) suggests a reliability of $r = +.90$ or higher for scores used for individual prediction purposes, and reliabilities of $r = +.70$ or better for prediction of group behavior. Reliabilities of less than $r = +.70$ are quite acceptable for basic research purposes. Some writers suggest that reliabilities of greater than $r = +.70$ are perhaps wasteful of measurement effort, because too much effort has to be given to measurement, rather than to showing relationships or lack thereof among variables. Reliability correlation coefficients closer to zero strongly suggest that observers or tests or test items do not agree among themselves, or that the human behavior being observed changes too radically over time to permit precise measurement.

Shaw (1983) studied the reliability of analytic essay scores over several writing samples. In his review of the literature, he pointed out that previous studies of the reliability of scores on essays (Kincaid, 1953; Godshalk et al., 1966; Breland and Gaynor, 1979) showed that a single essay scored by two readers had a reliability of from .38 to .60—not a strong correlation for placement—and a set of five essays had a reliability of .84. His study found similar results.

Influences on reliability

THE SPEARMAN–BROWN FORMULA: NUMBER OF RATERS OR ITEMS

The Spearman–Brown Formula helps researchers assess and increase the level of internal-consistency reliability in a test or measurement instrument. The formula (see Appendix I) shows the relationship between the length of the test (number of items) or the number of observers and the internal-consistency reliability (its precision of measurement). It also estimates the precision that will be achieved by increasing or decreasing the length of the test or the number of observers. This reliability (precision) or lack of it will markedly affect a researcher's ability to detect differences among groups and relationships among variables. Diederich (1974, pp. 33–34) discussed the use of the Spearman–Brown Formula to estimate higher reliability from essay tests.

He suggested that a researcher can more readily increase internal-consistency reliability by adding another essay test with two new raters
than by either increasing the length of time for an essay test or adding
more than two raters to one test.

INDIVIDUAL DIFFERENCES AND OTHER FACTORS

Reliability coefficients are correlations and, like all correlations, they
are subject to a number of influences. Reliabilities (and correlations) are
dependent on the degree of individual differences in the variables among
the people observed. For instance, the reliabilities of a test of vocabulary knowledge in a sample of college graduates will be less than the
reliability of that test in a general sample of adults, because the variation among scores in the college group will be less. It follows that the
correlations of this variable with other variables such as syntactic fluency
will also be less than in the general population. The limit on the degree
of correlation between two variables is a result of their reliabilities. There
is a direct relationship between the internal-consistency reliabilities of
the measurement of any two variables and the limits on the possible
observed correlation between the two variables (see Appendix I). The
higher the reliabilities, the higher the possible correlation. In fact, it is
difficult to interpret the real size and meaning of a correlation without
knowing the reliabilities of the measurement instruments involved,
particularly in composition research where the measured reliabilities of
the variables can be so diverse.

Other substantive factors influence reliability in composition studies:
(1) explicitness of the criteria, (2) amount and quality of the training
procedures (Coffman, 1971), (3) continual monitoring during the rating
(Cooper & Odell, 1977), (4) speed of rating (McColly, 1979), and (5)
background of the raters (McColly, 1979; Cooper, 1977).

Validity

The *validity* of a measurement system is its ability to measure whatever
it is intended to assess. Writing tests or ratings of writing are valid if
they have congruence with major components of writing behavior as
explained by theories in the profession. One of the recent issues in
writing assessment has been the challenge to the validity of objective
tests for measuring a productive skill such as writing. The Conference
on College Composition and Communication passed a resolution in
1978 that included the following statement: "No student shall be given

credit for a course, placed in a remedial writing course, exempted from a required writing course, or certified for competency without submitting a piece of written discourse" (Hoover, 1979). The issue of validity is also being raised on another basis. Some current composition theorists maintain that writing ability entails many arts, powers, and skills—inventional arts, audience-adaptive skills, flexibility in writing different types of discourse, and revising skills. In their judgment, therefore, measures of writing would be valid only if they were capable of taking these powers and skills into consideration (Odell, 1977, 1981). Thus, validity depends in important ways on social consensus.

Validity also rests on the soundness of interpretation of the measurement systems. Sometimes measurements are given far broader interpretations than their nature warrants. For example, the results of tests of writing done at one sitting (without time for planning or revising) and in one aim (expressive or informative) are sometimes misinterpreted to be valid tests of an individual's overall writing ability. Validity depends also on the extent to which tests or ratings predict other scores, later rating of behavior, or accomplishments that should theoretically be related to prior test scores and observations. Valid measures also differentiate among groups of people or students who theoretically should be different on certain variables: groups such as professional writers, freshman writers, and remedial writers.

While the question of reliability can be rather definitively answered, the question of validity can never be fully resolved. Writing theories continually grow, change, and sometimes contract. The validity of writing measurement instruments must continually be tested empirically in the light of new understandings in the field. The ethics of the use and interpretation of tests, ratings, and observations require that those who create and administer these measures must continually have evidence of their validity.

As a legal aspect of validity, if a measurement system is to be used for selection, placement, or graduation, and it reveals differences among ethnic, racial, or age groups, then unless clear, empirically established validity is present that test would be illegal under federal law. Because almost all tests show differences among at least one of these groups, it becomes clear that empirically established validities are necessary if an institution or school district wishes to avoid suits in federal court.

Types of validity

There are four major types of validity of measurement systems: content, concurrent, predictive, and construct.

CONTENT VALIDITY

Content validity is demonstrated by showing that the topics in a test are representative of the content for which the test is being used as a placement tool or a measure of achievement. Often independent, expert judgment is made about the congruence between the test and the behavior to be tested. Tests that exempt students from courses must therefore represent the content of those courses to have content validity.

CONCURRENT VALIDITY

Concurrent validity is the degree of relationship between a known, valid measure of behavior (rating systems, judgments, or tests) and a new measurement system. Typically, concurrent validity is used to establish the viability of a measurement system that is less costly or less time-consuming than available systems.

The methods used to establish this kind of validity are strictly empirical and usually correlational. The two measurement systems are applied to one or more groups independently, and the results are correlated. The two measurement procedures must be kept independent of one another so that there is no contamination or leakage of information between the two sets of measures. If the correlation is substantial and approaches its expected upper limit, considering the internal-consistency reliabilities of the test, then the new measurement system is considered to have concurrent validity with the well-established measurement system.

In 1977, the Educational Testing Service (ETS) correlated the essay test with the multiple-choice components of the English Composition Test and found a correlation of .50 and in 1978 of .48. The CLEP (College Level Examination Program) test of the ETS was originally validated during the 1940s by correlating its objective scores on usage and sentence correlation with five papers scored by five readers. The test achieved a .72 coefficient for concurrent validity. Given the reliabilities of these two tests, one can conclude that they are closely related. Most objective tests of writing have established concurrent validation by correlating multiple-choice tests with holistic essay scores. However, Quellmalz cites recent research that challenges the high correlations that these tests have received (Baker and Quellmalz, 1980). Charney points out the difficulty of finding a measure of writing whose validity is not in question to act as a criterion against which to measure another test (1984, p. 77).

PREDICTIVE VALIDITY

Predictive validity, as the name suggests, attempts to forecast or predict some future condition, score, or judgment. Prediction is a stringent criterion of a measurement system because it puts to trial the theory underlying a test or measurement system. Predictive validity, empirically shown on a regular basis against course grades, is imperative for tests that will be used for placement of students in remedial, beginning, or advanced English composition courses. It can also be useful for research purposes. In 1966, Godshalk et al. questioned grades as a criterion for predictive ability, but a study by Culpepper and Ramsdell (1982) confirmed other previous conclusions about the predictive validity of objective tests. Daly and McCroskey (1975) determined the predictive validity of the Writing Apprehension Test (WAT), finding that high apprehensives chose jobs that required less writing. The WAT also predicted placement (Daly and Miller, 1975b), academic choice (Daly and Shamo, 1978), and career choice (Daly and Shamo, 1976). Educational Testing Service also used predictive validity for the Standard Test of Written English, correlating the scores on this objective test with grades in freshman English classes.

CONSTRUCT VALIDITY

Construct validity is the measure's congruence with the theoretical concepts of a field. A construct in composition is a theory about the nature of writing behavior: composing processes, contexts, developmental levels, types of discourse, and so forth. When developing a new writing test, ETS, for example, engages expert writing consultants to list a set of specifications indicating the important components of writing to be tested. Their list is then sent to instructors throughout the country for response. On the basis of this survey, ETS constructs a test to assess as many of these powers and skills as they can. The writing component of the new National Teacher Examination, for example, now has an objective test of conventions and syntactic skills, an essay, and a third type of examination to assess those rhetorical skills that cannot be measured by either of the two other types of exams.

Some composition researchers consider holistic, analytic, or T-unit analyses of finished texts as measures too gross to assess such writing dimensions as planning, aim, audience adaptation, and revision (Quellmalz, 1981). Schuessler, Gere, and Abbott (1981) object to previous tests of attitude on the basis of lack of construct validity because these tests take into account only one or two dimensions of composition. Charney (1984) questions the construct validity of holistic scoring because of the variability of topics and types of discourse, the arbitrariness of the criteria, and the failure of raters to adhere to the criteria.

Construct validity, which is typically determined in a multivariate situation, has two components: discriminant and convergent. *Convergent validity* is the general agreement among ratings where measures should be theoretically related. These ratings are gathered by methods that are maximally independent of one another. *Discriminant validity* is a low relationship among measures that theoretically should not be related. Schuessler, Gere, and Abbott (1981) determined convergent validity by intercorrelating four scales—linguistic, maturity, student self-expression, standard English, and defining and evaluating writing tasks—establishing that items within each four scales were related to each other and that the scales were relatively independent of each other, thus not measuring a single unidimensional attitude. They also determined that their measure had discriminant validity, that it had little correlation with variables that the test was *not* supposed to measure. Table 7-1 shows Schuessler, Gere, and Abbott's construct validity.

To establish construct validity, a researcher can use the Campbell and Fiske (1959) multitrait–multimethod procedure, which correlates the new construct measure to be validated with other traits (multitraits) using at least two different methods (multimethods) in the data collection system. A reasonably large number of subjects must be used to be definitive. The correlations must show that each trait correlates with itself across measurement methods and that across traits the correlations are near zero except where they are theorized to be somewhat related. This latter requirement shows that the newly theorized construct is indeed a new one that does not overlap highly with prior variables.

Table 7-1 Intercorrelations of Scores on Four Scales Measuring Teacher Attitudes toward Instruction in Written Composition

	Standard English	Linguistic Maturity	Student self-expression
Linguistic maturity	.13	—	—
Student self-expression	−.47	−.00	—
Defining and evaluating writing tasks	−.07	.26	.06

Source: Scheussler, Gere, and Abbott (1981), Table 3. Copyright © 1981 by the National Council of Teachers of English. Reprinted by permission of the publisher.

The multitrait–multimethod approach to developing construct validity of variables also illustrates the principle of parsimony: the use of only one name to describe a variable or construct. If several variables intercorrelate about as high as their reliabilities will allow, it is both wasteful and confusing to have two or more names to describe that construct.

Another principle of science states that an abstract variable or construct should in great part be defined by identifying the operations by which it can be measured or the conditions that generate it. The field of composition would profit by having more operational definitions of each of its writing variables rather than primarily conceptual or subjective definitions. A theoretical definition is strengthened and clarified by operational definitions, which in turn are guided and enhanced by theory. If researchers want to define audience adaptiveness, for example, they can specify the features of audience-adaptive behavior that can be measured. The ultimate definition of a theoretical construct like audience adaptiveness can then be drawn from the multiple operational definitions that are necessary to define fully the construct. (This is impossible in any finite time, but is a goal toward which to strive.) The attempt clarifies the theory, and the clarified theory permits better operational definitions.

Validity and reliability

Validity and reliability influence each other in various ways. Measurement systems can be reliable without being valid, but if their validity is to be demonstrated, they must have some reliability. If the reliability of a measurement increases or decreases, it will influence the validity coefficients.

Summary

Measurement instruments are central to both descriptive and experimental research. Four levels of measurement are explained: nominal, ordinal, interval, and ratio, with nominal and interval constituting the major types used in composition research. Two requirements govern measurement—reliability and validity. Three major types of reliability are used: equivalency, stability, and internal consistency. The kind of analysis used to determine reliability is governed by whether the data are nominal or interval. Four kinds of validity can be determined: content, concurrent, predictive, and construct.

Checklist for Evaluating "Measurement Instruments"

1. What measurement instruments were used?
2. Were they standard ones or devised by the researcher?
3. Were the kinds of analyses used in the research appropriate for the data?
4. What were the kinds of reliability for these instruments? What levels were attained? What limiting effects did this reliability have on the research interpretation?
5. What kinds of validity did these instruments have?

References

Measurement Theory and References

Asher, J.W. (1976). *Educational research and evaluation methods*. Boston: Little, Brown.

Campbell, D.T., & Fiske, D.W. (1959). Convergent and discriminant validation by the multitrait–multimethod matrix. *Psychological Bulletin, 56,* 81–105.

Cohen, J. (1960). A coefficient of agreement for nominal scales. *Educational and Psychological Measurement, 20,* 37–46.

College Entrance Examination Board (1975). *The Test of Standard Written English: A preliminary report.* Princeton, N.J.: Educational Testing Service.

Cooper, C. (1987). *Writing assessment handbook, Grade 8* (Revised Edition). Sacramento: California State Department of Education.

Cronbach, L.J. (1971). *Test validation.* In R.L. Thorndike (Ed.). *Educational measurement.* Washington, D.C.: American Council in Education, pp. 443–507.

Diederich, P. (1974). *Measuring growth in English.* Champaign, Ill.: National Council of Teachers of English.

Edwards, A.L. (1970). *The measurement of personality traits by scales and inventories.* New York: Holt.

Hunt, K.W. (1965). *Grammatical structures written at three grade levels.* Urbana, Ill.: National Council of Teachers of English.

Jones, L.V. (1971). The nature of measurement. In R.L. Thorndike (Ed.). *Educational measurement.* Washington, D.C.: American Council in Education, pp. 335–355.

Kuder, G.F. & Richardson, M.W. (1937). The theory of estimation of test reliability. *Psychometrika, 2,* 151–160.

Likert, T.R. (1932). A technique for the measurement of attitudes. *Archives of Psychology,* No. 140, pp. 1–55.

Lloyd-Jones, R. (1977). Primary trait scoring. In C. Cooper & L. Odell (Eds.). *Evaluating writing.* Urbana, Ill., NCTE.

Lloyd-Jones, R. (1987). Tests of writing ability. In G. Tate (Ed.). *Teaching composition: Twelve bibliographic essays.* Fort Worth: Texas Christian University Press.

National Council of Teachers of English (1971). *Composition questionnaire.* Urbana, Ill.: NCTE Commission on Composition.

Nie, N.H. et al. (1979). *SPSS: Statistical package for social sciences,* 2nd edition. New York: McGraw-Hill.

Nunnally, J.C. (1978). *Psychometric theory*, 2nd edition. New York: McGraw-Hill.

Osgood, C., Suci, G., & Tannenbaum, P. (1969). Measurement of meaning. In J. Snider & C. Osgood (Eds.). *Semantic differential technique: A sourcebook.* Chicago: Aldine.

Polanyi, M. (1958). *Personal knowledge: Toward a postcritical philosophy.* New York: Harper & Row.

Scardamalia, M., & Bereiter, C. (1985). Research on written composition. In M.C. Wittrock (Ed.). *Handbook of research on teaching.* New York: Macmillan.

Schuessler, B., Gere, A., & Abbott, R. (1981). The development of scales measuring teacher attitudes toward instruction in written composition: A preliminary investigation. *RTE, 15,* 55–63.

Stanley, J. (1971). Reliability. In R.L. Thorndike (Ed.). *Educational measurement.* Washington, D.C.: American Council in Education, pp. 356–442.

Winer, B.J. (1971). *Statistical principles in experimental design*, 2nd edition. New York: McGraw-Hill, pp. 283–296.

Discussions of Writing Measurements

Breland, H.M., & Gaynor, J.L. (1979). A comparison of direct and indirect assessments of writing skill. *Journal of Educational Measurement, 16,* 119–128.

Breland, H.M., Camp, R., Jones, R.J., Morris, M.M., & Rock, D.A. (1987). *Assessing writing skills.* Research Monograph No. 11. New York: College Entrance Examination Board.

Bridgeford, N.J. (1981). *A directory of writing assessment consultants.* Portland, Or.: Clearinghouse for Applied Performance Testing, Northwest Regional Educational Laboratory.

Burt, F.D., & King, S. (Eds.) (1974). *Equivalency testing.* Urbana, Ill.: National Council of Teachers of English.

Connor, U. & Lauer, J.M. (1985). Understanding persuasive essay writing: Linguistic/rhetorical approach. *Text, 5,* 309–326a.

Connor, U. & Lauer, J.M. (1988). Crosscultural variation in persuasive student writing. In A.C. Purves (Ed.). *Contrastive rhetoric. Written communication annual,* Vol. 2. San Francisco: Sage Publications.

Cooper, C.R. (1975). Measuring growth in writing. *English Journal, 64,* 111–120.

Cooper, C.R. (1977). Holistic evaluation of writing. In C.R. Cooper & L. Odell. *Evaluating writing.* Urbana, Ill.: NCTE.

Cooper, C.R. (Ed.) (1981). *The nature and measurement of competence in English.* Urbana, Ill.: National Council of Teachers of English.

Cooper, C.R., & Odell, L. (Eds.) (1977). *Evaluating writing: Describing, measuring, judging.* Urbana, Ill.: National Council of Teachers of English.

Educational Research Service (1981). *Testing for college admissions: Trends and issues.* Arlington, VA.: Educational Research Service.

Fagan, W., Jensen, J., & Cooper, C.R. (1985). *Measures for research and evaluation in the English language arts.* Vol. 2. Urbana, Ill.: National Council of Teachers of English.

Fagan, W., Cooper, C.R., & Jensen, J. (1975). *Measures for research and evaluation in the English language arts,* Vol. 1. Urbana, Ill.: National Council of Teachers of English.

Faigley, L., Cherry, R., Jolliffe, D., Skinner, A. (1985). *Assessing writers' knowledge and processes of composing.* Norwood, N.J.: Ablex Publishing Company.

Godshalk, F.I., Swineford, F. & Coffman, W.E. (1966). *The measurement of writing ability.* New York: College Entrance Examination Board.

Greenberg, K., Weiner, H., & Donovan, R.A. (Eds.) (1986). *Writing assessment: Issues and strategies*. White Plains, N.Y.: Longman.

Grommon, A. (Ed.) (1976). *Reviews of selected published tests in English*. Urbana, Ill.: National Council of Teachers of English.

Hogan, T.P., & Mishler, C. (1980). Relationships between essay tests and objective tests of language skills for elementary school students. *Journal of Educational Measurement, 17*(3), 219–227.

Hoover, R. (1979). An annotated bibliography on testing. *CCC, 30*, 384–392.

Keepes, J.M., & Rechter, B. (1973). *English and its assessment*. Hawthorn, Australia: Australian Council for Educational Research.

Langer, J. (1984). The effects of available information on responses to school writing tasks. *RTE, 18*, 27–44.

Mellon, J. (1969). *Transformational sentence-combining*. NCTE Research Report No. 10. Champaign, Ill.: National Council of Teachers of English.

Meredith, V.H., & Williams, P.L. (1984). Issues in direct writing assessment: Problem identification and control. *Educational Measurement, 3*, 11–15, 35.

Mullis, I.V. (1984). Scoring direct writing assessments: What are the alternatives? *Educational Measurement, 3*, 16–18.

Murray, M. (1987). *Measuring insight in student writing*. Doctoral Dissertation, Purdue University.

Newell, G.E. & Mac Adam, P. (1987). Examining the source of writing problems: An instrument for measuring writers' topic-specific knowledge. *Written Communication, 9*, 156–174.

Odell, L. (1981). Defining and assessing competence in writing. In C. Cooper (Ed.). *The nature and measurement of competence in English*. Urbana, Ill.: National Council of Teachers of English.

Odell, L., & Cooper, C.R. (1980). Procedures for evaluating writing: Assumptions and needed research. *College English, 42*, 35–43.

Purves, A., et al. (1975). *Common sense and testing in English*. Report of the Task Force on Measurement and Evaluation in the Study of English. Urbana, Ill.: National Council of Teachers of English.

Quellmalz, E. (1984). Toward successful large-scale writing assessment: Where are we now? Where do we go from here? *Educational Measurement, 3*, 29–32.

Spandel, V., & Stiggins, R.J. (1981). *Direct measures of writing skills: Issues and applications* (revised edition). Portland, Oreg.: Northwest Regional Educational Laboratory.

Stiggins, R. (1982). A comparison of direct and indirect writing assessment methods. *RTE, 16*, 101–114.

White, E. (1985). *Teaching and assessing writing*. San Francisco: Jossey-Bass.

Wilkinson, A., Barnsley, G. Hanna, P. & Swan, M. (1980). *Assessing language development*. Oxford: Oxford University Press.

Wilkinson, A., et al. (1983). More comprehensive assessment of writing development. *Language Arts, 60*(7), 871–881.

Studies of the Reliability and Validity of Measurement Instruments

Ackerman, T.A. (1984). An investigation of the assessment of writing ability from a declarative/procedural perspective. *DAI, 45*, 12A.

Anderson, C.C. (1969). The new STEP essay test as a measure of composition ability. *Educational and Psychological Measurement, 20*, 95–102.

Asaad, S.B. (1984). The construction and validation of a multiple choice and a performance test of communicative writing ability for the EFL classroom. *DAI, 45,* 09A.

Baker, E.L., & Quellmalz, E.S. (1980). *Educational testing and evaluation.* Beverly Hills, Calif.: Sage Press.

Benson, G. (1979). The effect of immediate item feedback on the reliability and validity of verbal ability test scores. *DAI, 40,* 10A.

Biola, H.R. (1979). Performance variables in essay testing. *DAI, 40,* 10A.

Breland, H.M., et al. (1976). *A preliminary study of the Test of Standard Written English.* Princeton, N.J.: Educational Testing Service.

Breland, H.M., & Gaynor, J.L. (1979). A comparison of direct and indirect assessments of writing skill. *Journal of Educational Measurement, 16,* 119–128.

Charney, D. (1984). The validity of using holistic scoring to evaluate writing: A critical view. *RTE, 18,* 65–81.

Chase, C.I. (1968). The impact of some obvious variables on essay test scores. *Journal of Educational Measurement, 5,* 315–318.

Coffman, W.E. (1966). On the validity of essay tests of achievement. *Journal of Educational Measurement, 3,* 151–156.

Coffman, W.E. (1971). On the reliability of ratings of essay examinations in English. *RTE, 5,* 24–36.

College Entrance Examination Board (1975). *The Test of Standard Written English.* Princeton, N.J.: Educational Testing Service.

Crockenberg, S. (1972). Creativity tests: A boon or boondoggle for education? *Review of Educational Research, 42,* 27–45.

Culpepper, M.M. & Ramsdell, R. (1982). A comparison of a multiple choice and an essay test of writing skill. *RTE, 16,* 295–297.

Daly, J., & Dickson-Markham, F. (1982). Contrast effects in evaluating essays. *Journal of Educational Measurement, 19,* 309–316.

Daly, J., & McCroskey, J.C. (1975). Occupational desirability and choice as a function of communication apprehension. *Journal of Counseling Psychology, 22,* 309–313.

Daly, J., & Miller, M. (1975a). The empirical development of an instrument to measure writing apprehension. *RTE, 9,* 242–249.

Daly, J., & Miller, M. (1975b). Further studies of writing apprehension: SAT scores, success expectations, willingness to take advanced courses, and sex differences. *RTE, 9,* 250–256.

Daly, J., & Shamo, W. (1976). Writing apprehension and occupational choice. *Journal of Occupational Psychology, 49,* 55–56.

Daly, J., & Shamo, W. (1978). Academic decisions as a function of writing apprehension. *RTE, 12,* 119–126.

Follman, J.C., & Anderson, J.A. (1967). An investigation of the reliability of five procedures for grading English themes. *RTE, 1,* 190–200.

Freedman, S. (1978). Influences on the evaluators of student writing. *DAI, 38,* 09A.

Freedman, S. (1979). How characteristics of student essays influence teachers' evaluations. *Journal of Educational Psychology, 71,* 328–338.

Gere, A. (1980). Written composition: Toward a theory of evaluation. *College English, 42,* 44–58.

Godshalk, F.I., Swineford, F., Coffman, W.E. & Shaw, W.E. (1966). *The measurement of writing ability.* New York: College Entrance Examination Board.

Greenbaum, W. (1977). *Measuring educational progress: A study of the national assessment.* New York: McGraw-Hill.

Grobe, C. (1981). Syntactic maturity, mechanics, and vocabulary as predictors of quality ratings. *RTE, 15*, 75–85.

Hogan, T.P. (1980). Relationships between essay tests and objective tests of language skills for elementary school children. *Journal of Educational Measurement, 17*, 219–227.

Johnson, N. (1977). The uses of grammatical and rhetorical norms, pedagogical strategies, and statistical methods in designing and validating a composition placement instrument. *DAI, 37*, 05.

Kincaid, G.L. (1953). Some factors affecting variations in the quality of students' writing. *DAI, 13*, 05.

Kirrie, M. (1979). *The English Composition Test with Essay.* New York, NY: College Entrance Examination Board.

Markham, L.R. (1976). Influences of handwriting quality on teacher evaluation of written work. *American Educational Journal, 13*, 277–283.

McColly, W. (1979). What does educational research say about the judging of writing ability? *Journal of Educational Research, 64*, 148–156.

Michael, W., & Shaffer, P. (1978). The comparative validity of California State University and Colleges English Placement Test (CSUC-EPT) in the prediction of grades in a basic English composition course and of overall freshman-year grade-point average. *Educational and Psychological Measurement, 38*, 985–1001.

Modu, C.C. & Wimmers, E. (1981). The validity of the Advanced Placement English Language and Composition examination. *College English, 43*, 609–620.

Moslemi, M. (1975). The grading of creative writing essays. *RTE, 9*, 154–161.

Newcomb, J.S. (1977). The influence of readers on the holistic grading of essays. *DAI, 38*, 03A.

Nold, E.W. & Freedman, S.W. (1977). An analysis of readers' responses to essays. *RTE, 11*, 164–174.

Odell, L. (1977). Measuring changes in intellectual processes as one dimension of growth in writing. In C.R. Cooper, & L. Odell (Eds.). *Evaluating writing: Describing, measuring, judging.* Urbana, Ill.: National Council of Teachers of English.

O'Donnell, R.C. (1976). A critique of some indices of syntactic maturity. *RTE, 10*, 31–38.

Ozier, P.W. (1979). The concurrent validity of CLEP composition scores in relation to writing performance. *DAI, 40*, 09A.

Pollard-Gott, L., & Frase, L.T. (1985). Flexibility in writing style—a new discourse-level cloze test. *Written Communication, 2*(2), 107–128.

Quellmalz, E. (1980). *Controlling rater drift.* Los Angeles: University of California Center for the Study of Evaluation.

Quellmalz, E. (1981). Issues in designing instructional research: Examples from research on writing competence. Paper presented at the annual meeting of the American Psychological Association, Los Angeles.

Quellmalz, E., Cappel, F.J., & Chou, C. (1982). Effects of discourse and response mode on the measurement of writing competence. *Journal of Educational Measurement, 19*, 241–258.

Redfield, D.L., & Martray, C.R. (1984). *The prose quantification system: Development and construct validity.* Paper presented at the annual meeting of the American Psychological Association, Toronto, Canada.

Shaw, R. (1983). Stability of analytic essay scores: Implications for diagnosis and placement. Paper presented at the annual meeting of the American Psychological Association, Anaheim, Calif.

Stewart, M.F. & Grobe, C.H. (1979). Syntactic maturity, mechanics of writing, and teachers' quality ratings. *RTE, 13*, 207–215.

Torrance, P. (1968). Examples and rationales of test tasks for assessing creative abilities. *Journal of Creative Behavior, 2*, 165–178.

Witte, S., & Davis, A. (1980). The stability of T-unit length: A preliminary investigation. *RTE, 14*(1), 5–17.

Witte, S., & Davis, A. (1982). The stability of T-unit length in the written discourse of college freshmen: A second study. *RTE, 16*, 71–84.

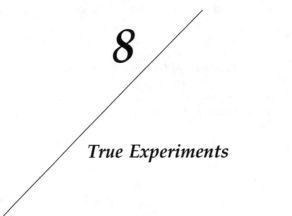

8

True Experiments

In the research designs discussed in the past chapters, investigators studied the world as they found it. These designs were descriptive; the researchers did not systematically control the conditions under which the various data were collected. They did not introduce treatments. They certainly may have taken advantage of naturally occurring changes or differences in conditions in the environment, to see if there were also changes or differences in the several variables being assessed. They may have looked for differences between males and females; among fourth-, eight-, and twelfth-grade students; between writers in remedial and regular composition courses; between academic and business readers-graders. They clearly did not, however, systematically change the variables under scrutiny—age, gender, occupation, or grade level. Unlike these descriptive designs, *experimental research* actively supplies treatments or conditions such as sentence-combining, planning strategies, or word processing to determine cause-and-effect relationships between treatments and later behavior. The basic principle here, as we stated in Chapter 1, is that no knowledge can be gained without comparison.

The *true experiment* (*true* is a technical term here; see Campbell and Stanley, 1963) is a special type of experimental design in which the researcher actively intervenes systematically to change or withhold or apply a treatment (or several) to determine what its effect is on criterion variables. The distinctive feature of true experiments is their use of randomization by which subjects are allocated to treatment and control groups. Treatments can also be randomly assigned to these randomized groups. In the simplest form of the true experiment, researchers present one treatment method to the experimental group and a traditional method to the control group. In a more advanced form, they use one treatment with one group, another treatment with a second group, and a traditional method with the control group of students. A true experiment is therefore one in which the subjects have been assigned randomly to the various groups, receive treatments or conditions, and then are evaluated on at least one measure in common.

True experiments can be used wherever instructors or program directors are making decisions about instructional methods. These decisions may be about whether or not to teach a certain planning strategy or revising skill, whether to use holistic or analytic evaluation, whether to use presentational, natural process, or environmental pedagogy or about any number of other instructional decisions. If instructors can use or reject any teaching methods, these methods can become treatments in a true experiment. Below is a brief description of three true experiments that we will use to explain this research design.

Three true experiments

Study 1   Frank O'Hare (1973). *Sentence-combining: Improving student writing without formal grammar instruction*. Urbana, Ill.: National Council of Teachers of English.

O'Hare randomly assigned eighty-three seventh graders to two treatment and two control groups. The treatment groups were taught a shortened version of the regular English curriculum and O'Hare's sentence-combining exercises without formal grammar instruction. The control groups were taught a regular English curriculum. O'Hare was interested in four questions (p. 2):

1. Would students in the treatment groups write syntactically different sentences from those written by students in the control groups?
2. If there were differences, would these be maturity differences?
3. Would students in the treatment group write better overall compositions?
4. What would be the curricular implications of these findings?

As a result of his analysis, he concluded that sentence-combining *caused* the positive changes hypothesized in his questions. We will examine this study in more detail as we explain the techniques of true experiments.

Study 2   Jeanne Halpern and Sarah Liggett (1984). *Computers and composing: How the new technologies are changing writing.* Carbondale, Ill.: Southern Illinois Univ. Press.

In 1982, Halpern and Liggett conducted a true experiment to test the effect of teaching dictation strategies to business writing students. They theorized that students taught both the composing and technical processes of dictation would dictate better memos than would students taught only the technical processes. Eighty-four students were randomly assigned to four classes taught by two instructors—two experimental and two control classes. (Each instructor taught one experimental and one control group.) The experimental group did significantly better than the control group. We will analyze this study in some detail below.

Study 3   George Hillocks, Elizabeth Kahn, and Larry Johannessen (1983). Teaching defining strategies as a mode of inquiry. *RTE, 17,* 275–284.

Hillocks, Kahn, and Johannessen were interested in the effects of teaching strategies of defining. They theorized that when students practiced strategies of defining, they would apply them to new concepts, giving their definitions more numerous and appropriate criteria, examples, and clarifying contrasts than those of students whose instruction included only explanations of definition, analysis of models, and practice. The experiment supported these theories.

Research design: true experiments

Hypotheses

When researchers report experiments in journals or books, they state their expectations of the results of treatments on criterion variables in the form of hypotheses. Many composition researchers phrase their hypotheses in the form of questions or positive statements.

O'Hare stated his hypotheses positively.

1. The experimental group, which was exposed to the sentence-combining practice, will score significantly higher on the six factors of syntactic maturity than the control group, which was not exposed to the sentence-combining practice.
2. The experimental group will write compositions that will be judged

by eight experienced English teachers as significantly superior in overall quality to the compositions written by the control group.

Hillocks, Kahn, and Johannessen also used the positive hypothesis form.

1. When students practice strategies of defining, they will apply them to new concepts, making their definitions include more numerous and appropriate criteria, examples, and clarifying contrasts than those of students whose instruction includes only explanations of definition, analysis of models, and practice.
2. The control students will also make some improvement.

Formal statistical analysis requires that the hypotheses be cast in the null form. A *null hypothesis* states that no difference between the groups will result from either the treatment or the control situation. Strictly speaking, then, all the researcher has to do is reject or accept this null hypothesis via the statistical analysis. As a result of randomization, the researcher expects no differences between groups. At least, the researcher expects that the treatment will do no harm; to prove the efficacy of a treatment, therefore, the researcher need only reject the null hypothesis. Witte and Davis (1982), in their study of T-unit stability, discussed in the previous chapter, used formal, statistical null hypotheses.

1. There will be no more variability in mean T-unit length between essays of the same type written by different students than between essays written by the same student.
2. There will be no significant difference between the mean T-unit length of informative essays employing classification and informative essays using comparison when all essays are written by the same group of students (pp. 78–79).

Randomization

In experimental studies, researchers use randomization to establish groups that can be considered "not unequal." If subjects are randomly allocated to two or more groups, the researcher can generalize that, over a large number of random allocations, all the groups of subjects will be expected to be identical on all variables. In other words, when large numbers of students or classes of students are randomized to several instructional methods groups, all variables associated with those students or classes are also allocated to several groups strictly by chance. Given this situation, then, the researcher can state that there are no expected differences in the long run among the students in the several instructional groups on *any* variable. This equivalency is true for all variables (age, writing skills, sex, descendants of Shakespeare) *except as would be expected by chance.* Chance differences can be well defined quantitatively. Further, all groups are expected essentially to be the same both at the time of the randomization and for all future time unless there is an intervention that acts differently on one or more of the groups. For researchers using behavioral and social science methodologies, there exists, then, a powerful research logic: when "not unequal" groups are acted on by unequal treatments, unequal groups result.

In the actual research, however, the logic works in reverse. Researchers develop groups expected to be equal through randomization, apply treatments of unknown or unproven efficacy, and thereby learn whether these treatments are effective by examining the results. The method of randomizing is the same as discussed in Chapter 4 on sampling. The most common way is to assign each student or class a number and then, using the table of random numbers, assign systematically by prearrangement every student or class to one of the two or several groups—treatment or control.

Where a large number of classes are being taught much the same material, such as in freshman composition courses, it is possible to allocate whole classes randomly, with their instructors, to several experimental and control groups. The generalizability is markedly enhanced because far more instructors are involved, and they too are then considered "not unequal" across instructional groups. The number of observations, however, is smaller for the statistical analysis. (See discussion of degrees of freedom in Appendix I.) This does not cause the analysis to lack definitiveness because class means, based on 15 or 20 students, are far more stable than observations or test scores on individuals.

All three studies described here used randomization. With the cooperation of the school administration, O'Hare randomized 83 seventh graders to two treatment and two control groups. Halpern and Liggett planned their experiment, with the aid of the English department schedule deputy, six months in advance in order to schedule four sections of business writing (eighty-four students) into two time slots. After meeting all students in one room at each of the two hours, they randomly assigned them, using a table of random numbers, to one treatment and one control group (within each of the time slots). Each instructor taught one treatment and one control group alternatively in each of the two time slots. Randomizing the students gave the statistical analysis 84 observation units. If the randomization had been by the intact classes, only four observation units would have been the basis for the statistical analysis. Hillocks, Kahn, and Johannessen randomized twelfth-grade students into two instructional groups by coin tossing, a random process. They also added a nonrandomized group from an eleventh grade at another school. For all subjects they added a pretest, which was a paper of definition written in two 50-minute class periods.

Pretests

Randomization precludes the necessity of pretests to demonstrate the equality of the groups. Campbell and Stanley indicate that a pretest design is not quite as good as the nonpretest randomized design (1963, p. 226). Sometimes, however, a pretest can be helpful. When a subject drops out, leaves, or refuses to write the composition, the random groups are no longer strictly equal. The researcher will not have any criterion data on such students. Students often do not drop out at random, having been selectively "pushed out" by the treatment. Statistical tests of a null hypothesis of no difference between groups (see above) are less realistic because the selective elimination may cause differences. The subjects remaining in the groups will differ as the result of these dropouts. Pretests will help the researcher to determine the characteristics of those who left and their possible influences on the criterion variables. Pretest data are also useful in increasing the power of statistical analyses, such as correlated data *t*-tests or repeated measurements analysis of variance, so that a researcher reduces the chances of failing to detect true differences.

O'Hare added a composition pretest, which proved to be helpful because one student from a control group and one from a treatment group dropped out. With pretest scores, he could determine the probable influence of these dropouts on the posttest data and still draw sound conclusions. O'Hare's study also inadvertently provided an empirical example of the fact that randomization does indeed create equal groups. Table 3 in his report (p. 53) compared the pretest mean scores on the six factors of syntactic maturity between the randomized experimental and control groups. They were, respectively: words per T-unit, 9.63 and 9.69; clauses per T-unit, 1.36 and 1.37; words per clause, 7.06 and 7.05; noun clauses per 100 T-units, 13.76 and 13.67; adverb clauses per 100 T-units, 14.34 and 14.24; and adjective clauses per 100 T-units, 7.90 and 9.29. (None of these differences was significant, the largest *t*-test value being 1.16.) Precisely half of the variables' *t*-tests were negative and half positive; in other words, for half the variables the control groups' means were numerically larger and for the other half the experimental groups' means in the sample were numerically larger. With samples this size and reasonable standard deviations (indicating the range of abilities), it was unlikely that true differences between the groups would have been undetected by the statistical analysis. The equality of the groups was exactly as would be expected from random allocation in reasonably sized groups.

A composition researcher would profit both in time and effort by taking seriously these clear benefits of random allocation. However, in this early study, O'Hare did not seem to trust randomization fully when he stated: "There was no practical way to control for the language experiences of the subjects outside their English classrooms" (p. 35). He also checked the number of books read by the students and queried other teachers about the students' assignments, finding in each class no difference between the treatment and control groups. We note that these efforts were unnecessary because O'Hare's randomization had already been expected to equalize these variables.

In much composition research, it may not be worth the effort to read all those pretest essays or count all those T-units to develop some variables that moderately correlate with the criteria. Halpern and Liggett's design saved the costly and time-consuming need for pretreatment scores, which would have required a number of experts reading a large number of papers. The true experiment has a design that looks like the pattern in Figure 8-1.

Treatments

In a true experiment in composition studies, the treatment is often an instructional strategy, a classroom environment, or some condition in-

where: R = a random group
       X = a treatment or condition
       O = an observation or measurement

Figure 8-1 Schemata for a true experiment.

troduced by the experimenter with the treatment group. The control group acts as a standard, typical condition, a foil in which the special treatment is absent. Sometimes a researcher has two or more different treatments or combinations of treatments and a control.

Treatments are often derived from new theories about writing that take better account of the nature of composing, the learning process, and the backgrounds and needs of students. They also represent classroom practices that seem intuitively effective. In the research report, whether in journal or book form, treatments and hypotheses (see below) are usually explained in the context of a literature review that reveals significant problems and questions that need experimental testing. A field like composition is rife with treatments being administered everyday to myriads of students. More recently, instructors have begun to examine these pedagogies both theoretically and empirically, instead of just anecdotally, in the journals.

O'Hare's treatment was the use of sentence-combining without formal grammatical instruction in conjunction with a short version of a regular English curriculum. The control group was taught the regular English curriculum. Halpern and Liggett's treatment was about four hours of instruction in strategies for effective dictation (SED), which gave students help with the composing process of dictation. Both the treatment and control groups met together to learn the technical aspects of dictation—using the equipment and addressing the transcribers. In the Hillocks, Kahn, and Johannessen study, both the treatment and control groups were taught the nature of definition through textbook explanations, analyses of models, and practice in writing and evaluating definitions. In addition, the treatment group received instruction in *strategies* of defining.

When a sole group is given a treatment without a control group, even though pretests and posttests are administered, little cause-and-effect knowledge can be gained. Such a research design is called a *preexperiment*. It is essentially exploratory. While it might be categorized as quantitative descriptive research because statistical analyses are used to show relationships, it differs from quantitative descriptive research because a treatment has been introduced by the researchers followed by a lapse of time before the criteria are observed.

Criterion variables and measurement instruments

Criterion variables are those used to detect differences between groups after the treatment has been administered. To assess these variables, researchers use a variety of measurement instruments like holistic scoring, counting, or scales of various kinds. O'Hare had seven criterion variables: words per T-unit, clauses per T-unit, words per clause, noun clauses per 100 T-units, adverb clauses per 100 T-units, adjective clauses per 100 T-units, and a single qualitative holistic judgment based on the factors of ideas, organization, style, vocabulary, and sentence structure. (The last criterion was simply scored 0 or 1.) For each of these operationally defined variables, O'Hare provided conceptual definitions from the literature on the construct of syntactic maturity. His analyses included calculations of the means and standard deviations of the six criterion variables and comparisons among groups, followed by analyses of variance between scores of pretests and posttests. He used eight experienced English teachers to select the best paper in matched pairs of thirty narrative and thirty descriptive compositions written by students paired by sex and IQ. He analyzed these frequency data using Chi-square analysis.

Halpern and Liggett had two trained raters use holistic grading and an analytic scale for their seven criterion variables of a dictated first-time final memo: heading, purpose, statement, summary, body, format, grammar and mechanics, and style. Each rater assessed all papers blindly, in other words without knowing which students were in the treatment or control group, thus without a possible bias. The reliability of measurement of the criterion variables ranged from correlation coefficients of .25 for grammar and mechanics to .82 for the holistic scoring to .90 for the summary. They used the F-test to assess differences between the two groups on each intervally measured criterion variable.

Hillocks, Kahn, and Johannessen had five criterion variables: quality of the class, quality of the differentia, number and quality of the criterion, number and quality of the elaborate example, and number and quality of contrastive examples. The three raters, who had been trained for eight hours, used a point scale for each of these variables (p. 280). The raters' reliability correlations (on a sample of fifteen papers) were .92, .93, and .94.

PROBLEMS WITH MEASUREMENT INSTRUMENTS IN TRUE EXPERIMENTS

As we have said in the previous chapter, composition research has often relied on measurement instruments that assess products, for example holistic, analytic, and T-unit analyses. Most true experiments have therefore tended to study the effects of treatments for which there are criterion variables that can be measured readily by existing instruments. But problems have arisen in the cases in which true experiments have been designed to study the effects of instruction on such process-oriented facets as invention, inquiry, cognitive development, audience adaptiveness, or revising. When researchers have used product measures on criterion variables such as revising or inquiry skills, these measures have often been inadequate, invalid, and too gross to detect significant differences on these composition variables (Quellmalz, 1981). Some researchers are beginning to create more finely tuned instruments to measure such variables. Researchers like Odell (1981), Young and Koen (1973), and Bereiter and Scardamalia (1983) have been working with ways to assess the growth of such skills as problem formulation, exploratory behavior, goal setting, large-scale revising, audience adaptiveness, and organizing. Some are using interviews, videotapes, and pictures as criterion variables in true experiments.

Statistical significance

Researchers hope that the treatment group will show statistically significant differences on the criterion variables tested. But even though true experiments have the greatest power to show cause-and-effect relationships, they never lead to certainty. Statistical analyses test hypotheses rigorously, but their results are *probabilistic*, just as all knowledge in science is probabilistic. The behavioral and social sciences are simply more so. Any variable can always differ between groups by chance alone. The principle is usually stated thus:

> As a result of random allocation of subjects into groups, all variables associated with the subjects will not be different among the groups except *as would be expected by chance.*

What does "except by chance" mean? Chance differences can be rather precisely defined. They are proportional to the individual differences in variables as expressed by their standard deviations and inversely proportional to the sample sizes (see Appendix I). Because chance results are possible, two kinds of statistical errors can occur.

TYPE I ERROR

A Type I error occurs when a researcher calls a result statistically significant that, in fact, is not so, but instead is due to chance. When a statistically significant difference occurs by chance alone, the treatment has really made no difference; its results, however, are interpreted as statistically significant. Type I errors are likely to occur only one time in a hundred or one time in twenty, depending on whether the $p=.01$ or the $p=.05$ level of significance is used. After several studies are done, the effectiveness or ineffectiveness of the treatment will become clearer because chance differences will balance out and true differences will become more apparent. Over even two studies, the chance is only one in 10,000 that two truly "no different" results on the same variable will be significant two times in a row at the $p=.01$ level. Researchers especially cautious to avoid Type I errors should set the p level low, at $p=.01$, where results that are improbable are expected to happen only one time in 100.

TYPE II ERRORS

Type II errors occur when researchers are unable to detect statistically significant differences between groups when in fact they exist. Type II errors pose serious problems for composition research because they may lead researchers not to recommend highly valuable teaching strategies that really did produce differences. This type of error occurs when human variability, imprecise measurement systems, and small numbers of subjects combine to produce large natural variabilities among groups. Under these circumstances, differences between the groups produced by the treatment are difficult to detect, even though they exist in the population. To avoid Type II errors, researchers can set the p values higher than the traditional .05 and .01, for example at .10 or .20. These are very respectable values, especially in introductory research when an investigator has a small sample. Other ways to help avoid Type II errors are to increase the sample size and the reliability of the measures by adding items to a test or increasing the number of observers doing ratings.

SELECTING p VALUES

As we have said, the selection of a p value should be based on the status of current knowledge in a field and on the cost of being wrong in one's conclusions. Because composition researchers should be concerned about losing useful methods of instruction or possibly overlooking valuable theory, a case can be made for setting p values at $p=.10$ and even $p=.20$. If one relationship in ten or in twenty is declared significant when it is not, the risk of error is far less than occurs now when few teaching methods are empirically tested. Further, using higher p values is no longer difficult because computer analyses report exact p

values. If researchers set higher p levels, in other words $p = .10$ or $p = .20$, better instructional methods are not likely to be lost.

Effect size

Effect size of the treatment over a series of studies is quickly becoming as much the decision rule of importance as statistical significance. Effect size is calculated by dividing the differences between means of data from the experimental and control groups by the combined standard deviation of the control and experimental groups. Another way of calculating effect size is to use the square root of the mean square used for the error term in a simple one-way analysis of variance. See Glass, McGaw, and Smith (1981 Chapter 5) for additional rules. A standard score, z, is derived that shows the effect size, the strength of the treatment group effect as compared with the control group effect. Effect sizes range from about -3.00 to 0 to $+3.00$. An effect size of about .20 to .50 is important; anything above .50 is a major difference. Effect sizes for studies can be related to major variables in experimental and control groups. The concept of effect size is already strongly influencing theory building in psychology and education where nonsignificant results over a series of experiments are likely to be prevalent, given the conditions of behavioral science research. Statistical significance is only too tenuously related to the importance of a variable in theory.

Results

Table 8-1 presents O'Hare's posttest means and standard deviations of the six criterion variables for all the groups. (These should always be reported.) We calculated O'Hare's effect sizes by subtracting the pretest means from the posttest means of the experimental and control groups, which in turn were divided by the combined standard deviations.

- For words per T-unit, $(15.75 - 9.96)/2.41 = 2.40$.
- For clauses per T-unit $(1.84 - 1.41)/.22 = 1.96$.
- For words per clauses $(8.55 - 7.03)/.89 = 1.71$.
- For noun clauses per 100 T-units $(23.55 - 15.85)/10.21 = .76$.
- For adverb clauses per 100 T-units $(29.01 - 15.50)/9.55 = 1.42$.
- For adjective clauses per 100 T-units $(31.61 - 10.16)/11.47 = 1.87$.

Table 8-1 Comparison of Posttreatment Mean Scores on the Six Factors of Syntactic Maturity: Experimental and Control Groups

Factors	Experimental (N = 41)		Control		t-Value	df
	Mean	SD	Mean	SD		
Words/T-Unit	15.75	3.00	9.96	1.64	10.88***	62
Clauses/T-Unit	1.84	.27	1.41	.16	8.72***	64
Words/Clause	8.55	1.02	7.03	.75	7.72***	73
Noun Clauses/ 100 T-Units	23.55	11.93	15.85	8.2	3.42***	71
Adverb Clauses/ 100 T-Units	29.01	11.15	15.5	7.67	6.41***	71
Adjective Clauses/ 100 T-Units	31.61	15.21	10.16	5.86	8.44***	51

Source: O'Hare (1973), Table 6. Copyright © 1973 by the National Council of Teachers of English. Reprinted by permission of the publisher.

t-test for two independent samples assuming unequal variances.
***significant at or beyond the .001 level.

It is clear that on all the criterion variables, sentence-combining produced major effect sizes. The strongest effect size was for words/T-Unit—the average student in the experimental group surpassed most students in the control group.

Halpern and Liggett also achieved significant results. On their holistic scoring, the experimental group's mean was 2.24; the control group's mean was 2.03 (based on a scoring system of 1 to 4). The standard deviations were .74 and .81; the difference between the means of the two groups was .21. Using the control group standard deviation of .81 as a reference (the older method of calculating effect sizes used by Glass, McGaw, and Smith, 1981), they calculated the effect size to be about .25—meaning that roughly the average student in the experimental group did better than 60% of the control group. (Note that by chance alone, one would expect the mean of the experimental group to exceed about 50% of the control group subjects.) On three of the criteria, heading, purpose statement, and format, the treatment group was significantly better. When all seven criteria were added together for the total analytic score, the experimental group was superior: the mean of the experimental group was 16.13; the mean of the control group was 14.37. The difference of 1.76, compared to the control group's standard deviation of 2.97, produced an effect size of .59. Thus the average student in the experimental group did better than about 72% of the students in the control group. The last four criterion variables were declared "not significant" because they were obscured by the individual differences among students and the imprecise measurement procedures. (They did not use a very large number of raters.) All four of the means in the experimental group, however, were larger than the means in the control group (a result that alone was somewhat improbable by chance). On the basis of these results and the research design, Halpern and Liggett could conclude that the treatment caused improvement in the quality of the dictated memos.

In the Hillocks, Kahn, and Johannessen study, Table 8-2 shows the results of a one-tailed t-test. The differences between the pretest and posttest of 10.35 and 8.05 points for the experimental group were significant at the $p < .0005$ level. The mean gain difference between the experimental Group A (the one that had been randomized with the control group) and the nonrandomized intact control group was 3.64, significant at the $p = .01$ level. The researchers also calculated effect sizes of 2.48 for the experimental Group A and .65 for the control Group A.

Internal validity

As we have said, one of the major aims in empirical research is to establish cause-and-effect relationships and the strength (effect size) of these relationships. In general, the ability of an investigator to infer possible cause-and-effect relationships in research designs is based on the quality of the design's internal validity. But several circumstances can flaw the cause-and-effect conclusiveness of research designs. Such flaws are called *threats to internal validity*. True experiments are likely to

Table 8-2   Pretest and Posttest Cumulative Scale Scores

	Experimental A ($n = 13$)	Experimental B ($n = 21$)	Control A ($n = 15$)
Pre-test Mean	4.20	6.74	7.75
SD	1.97	3.97	5.21
Post-test Mean	14.55	14.79	10.46
SD	5.48	5.13	3.86
Mean Gain	10.35	8.05	2.71
t Value	−6.49	−6.23	−1.92
Significance	$p < .0005$	$p < .0005$	$p < .05$
Effect Size	2.48	1.93	.65

Source: Hillocks, Kahn, and Johannessen (1983), Table 1. Copyright © 1983 by the National Council of Teachers of English. Reprinted by permission of the publisher.

have the strongest internal validity of any research method because randomization has controlled most of the threats to internal validity.

Measurement threats to internal validity

Instrumentation, Testing, Regression-Toward-the Mean, and Reactivity are the major threats to internal validity.

INSTRUMENTATION

Instrumentation involves changes in the measurement system such as bias, the lack of perfect calibration of alternative forms of measurement instruments, or differences among observers' ratings on the same dimension of behavior. It is risky to say, for example, that the writing ability of one group of students is greater than another's because the first group got higher grades in their English courses. Researchers who compare grades of differing nonrandomized, intact groups of students will often make erroneous interpretations with respect to Instrumentation, because teachers in one school or university may use widely divergent grading standards from those in another school. Randomization, however, controls the threat of Instrumentation. If a group of teachers in a school or raters in the same study are divided at random into two or more groups, these teachers' and raters' judgments on compositions can be expected to be equal along with all the other variables in the study. As a result, the threat of Instrumentation will not be a concern. Instrumentation was not a threat to the internal validity of the three studies discussed in this chapter because of randomization and the fact that only one group of raters assessed the compositions.

TESTING

The threat of Testing is the result of giving subjects the same or similar sets of tests more than once. This occurs frequently when pretests are used also as posttests. Testing threatens the internal validity of research designs using nonrandomized, intact groups by possibly increasing the inequality of the groups through pretesting. In true experiments, randomization controls this threat. If O'Hare's pretesting caused slight gains among the six criteria, it had similar effects on both randomized groups. Hillocks, Kahn, and Johannessen might also have risked the threat of Testing by using the same topics for both the pretest and posttest. The threat of Testing did not occur, however, because they did this for the randomized control and experimental groups and indicated that students did not know they would receive the same topics for the posttest.

The literature on research design has overemphasized, however, the Testing threat to internal validity. Lana (1959) concludes on the basis of his research on this threat that pretest sensitization is probably less of a threat than was previously feared. Willson and Putnam (1982) summarized the results of 32 studies investigating pretest sensitization effects of Testing. The overall effect size was a z score of +.22, indicating the general effect of using pretests. These effects differed, however, by the psychological area being assessed: for cognitive science, a z score of +.43; for attitudes, a z score of +.29; for personality, a z score of +.18; and zero for other areas. If the time between pretesting and posttesting is small (one day or less) or rather large (over one month), the effect size is small. Sameness or differences between pretest and posttest are not related to effect size.

REGRESSION-TOWARD-THE-MEAN

Regression-Toward-the-Mean is a major threat to the internal validity of research throughout the behavioral and social sciences. Because of the principle of imprecise measurement, the observed score (what the researcher finds) is always expected to be farther from the mean than the subject's true score. The differences between the observed and expected scores are negligible for persons already close to the mean on highly reliable instruments. But these differences are not negligible for persons at the extremes of the populations, for example remedial and honors students.

The degree of Regression-Toward-the-Mean expected is a function of the distance of the scores from the mean and the reliability of the measures. For instance, a person with a score two standard deviations from the mean, $z = +2.00$ on a measure with a reliability of $r = .70$ would have an expected z score of $+1.40$ ($12.00 \times .70 = 1.40$) on another measure of the same variable. This difference of .6 of a standard deviation is a rather major one, based simply on the inability to measure the variable with more than moderate precision. This Regression-Toward-the-Mean phenomenon works at the low end of the measurement scale as well. A z score of -2.00 with a reliability of $r = .70$ also produces a Regression-Toward-the-Mean effect of .6. Because reliability of measurement tends to be reduced over time, Regression-Toward-the-Mean becomes more pronounced over time.

In the three studies discussed here, randomization has equalized the

Regression-Toward-the-Mean threat in both the experimental and control groups on all measures used. Further, because these studies used typical populations—seventh-grade, twelfth-grade, and college populations—the groups would not be expected to regress toward the mean. In O'Hare's research, it was likely that the highest-scoring students on his pretests in both his groups would drop a little in their scores on all of his six criteria in posttesting. But this effect was counterbalanced by the pretesting effect, the maturation of the students over the course of the semester, and the fact that his lowest-scoring students on the pretest would also regress toward the mean on the posttest. Composition researchers who are interested in expert, novice, gifted, and remedial students must take care to control this threat. Fortunately, in true experiments, randomization allocates subjects so identified into both groups equally.

Other threats to internal validity

In addition to the measurement-related threats to internal validity of research, a number of additional threats arise from the nature of subjects and the factors surrounding the research. Some of these are Mortality, Maturation, Selection, History, Instability, and Social-psychological. Randomization also helps to control these threats except for History, Instability, and the Social-psychological.

MORTALITY

Mortality is any reduction in the experimental and control groups. In true experiments this can occur after randomization has taken place because students drop out of the study. If they do so because they do not like the instructional method, then dropping out becomes a criterion variable that is influenced by the treatment. (Interestingly, the reverse of dropping out may also occur. A particular method of instruction may attract students, also incurring the threat of Mortality.) In a true experiment, the threat of Mortality may cause problems of interpretation. Even though a researcher can ascribe differences in retention (which is a criterion variable in itself) to the effects of the treatments, the logic of the true experiment may be flawed because of the difficulty of obtaining observations or test scores on those who left the program. It is possible that only the better students remained in the program, achieving higher scores even without the treatment.

Mortality was not a problem in the Halpern and Liggett study. In the O'Hare study, only two students, one from each group, left the study because they withdrew from the school. Only one student entered but the data were entered into the analysis. In the study by Hillocks and colleagues, no report was given on the retention rate of students. In statistical analysis, ways have been developed to address the problem of missing data. This topic is treated in courses in statistics.

MATURATION

Students do grow and learn on their own. The Maturation threat to internal validity includes both biological and psychological variables. When interpreting data, a researcher must determine what students might have learned as a result of everyday growth and experiences. Maturation effects can be a problem in remedial elementary programs, because gains in learning or quality of writing may in part be due simply to maturation and increased cognitive development. In high school and college, the total educational program may have a maturing influence on the ability to write. Improvement may be mistakenly attributed to a remedial program. Here again, in the three studies, randomization has distributed any such influences equally.

SELECTION

Selection, an obvious threat to internal validity, occurs when a researcher claims that a treatment made a difference in groups that differed before the treatment was given. Such groups should differ similarly on many criteria after the treatments have been given. Many schools claim credit for improving their students' writing when, in fact, the students may have been good writers before instruction was given. The true experimental design in the three studies examined here controlled the Selection threat by equalizing the groups before the treatment began.

HISTORY

The History threat occurs when surrounding external conditions cause either members of the control or experimental group or both to change in ways that they ordinarily would not. A winning football team in a school can redirect the attention of the students, or a fire drill or a heat problem can influence behavior. Faculty changes can affect goals and achievement levels. The subject matter of compositions may affect performance. Both experimental and treatment groups are usually subjected to the same large external events. But sometimes the illness of an instructor or the loss of air conditioning may affect only the experimental or the control group. In this case History becomes a threat to a true experiment.

In the O'Hare and Halpern and Liggett research, History did not seem to pose a threat to either the T-unit and clause length counts or the dictation scores of the college students. O'Hare made valiant attempts to control for the threat of History from the influence of outside language factors (p. 38), but such efforts were somewhat unnecessary because the use of randomization brought outside influences under control.

INSTABILITY

The threat of Instability is the sum total of many factors in behavioral science research such as the fluctuation of human variables over time, the imprecise measurement systems, and the natural variations in classroom instruction. The result is a somewhat increased inability to detect true differences between experimental and control groups in the statistical analysis or the chance occurrence of false differences. Interestingly, Instability is the only threat to internal validity that is controlled by statistical analysis.

The threat of Instability may be one reason why Halpern and Liggett did not find significant differences for four criterion variables: summary, body, grammar and mechanics, and style. Two of these variables had the lowest interrater reliabilities, thus possibly preventing the researchers from detecting differences between the treatment and control groups.

SOCIAL-PSYCHOLOGICAL THREATS

In recent years, four social-psychological threats to internal validity have been suggested: Diffusion, Compensatory Equalization of Treatments, Compensatory Rivalry, and Resentful Demoralization. *Diffusion,* at one time called "leakage," occurs when participants of one group communicate with those of another group. Diffusion can be reduced by lowering the visibility of the differences between treatment and control groups and by urging teachers to avoid highlighting the variations among instructional methods. Using quasi-experiments (see Chapter 9) or doing the research at widely different sites can also reduce Diffusion. *Compensatory Equalization of Treatments* occurs when administrators or others demand that the "benefits" of the treatment be extended to the control group. *Compensatory Rivalry* and *Resentful Demoralization* occur when subjects have emotional responses to the experimental or control conditions. Randomization may help to avoid these last three threats by making the allocations of treatments fairer as the result of chance.

In the O'Hare study, because the gains of the experimental group were so much higher than the control group, these threats probably were not of consequence in the interpretation of the results. In the Halpern and Liggett study, most of the 15 weeks of the course was the same for both groups, including a common presentation of the technical process of dictation. Further, neither group was aware that the other group's instruction differed. In the Hillocks and co-workers study, the treatment was administered at two different schools, thus making this partially a quasi-experiment and less subject to these social-psychological threats.

INTERACTION EFFECTS AMONG THE INTERNAL VALIDITY THREATS

As indicated earlier, an interaction occurs when two or more factors combine to create greater or lesser scores on criterion variables than would be expected by a simple summing of their effects. For example, sex and grade level may create an interaction on T-unit length. A researcher would expect increases in T-unit length with grade level; similarly, girls would be expected to write longer T-units than boys. But if girls continue to improve over the grades at a *faster* rate than boys, then an interaction of grade and sex has occurred. In other words, the simple additive effects of grade and sex would have been surpassed. It is possible also that all the threats to internal validity can interact. With randomization, however, all interactions usually are likely to be equalized among the groups if the experimental and control instruction and the measurement are given simultaneously. In the three studies, no interactions of the threats to internal validity seem to have compromised the interpretation of cause-and-effect results.

External validity

External validity refers to the generalizability of research results, the power to extend the causes and effects of one experiment to other groups, other treatments, and other criteria like the ones in a given experiment. First, it should be clear that the external validity of a study can be no better than its internal validity; but because external validity is an inductive inference, it can never be completely established.

Threats to external validity

The external validity question for the O'Hare study concerned whether sentence-combining, which was effective with seventh graders of an average IQ of 112, would be effective with other seventh graders like his, with other degrees of intellectual maturity, or with other levels of students like eighth or twelfth graders. To answer this question, one must essentially hypothesize no interaction effects between instruction treatments and grade levels or ability levels for the populations to which the external-validity generalization is being applied. If one can accept the hypothesis, then one can generalize. If, on the other hand, one speculates that there are interaction effects, the next step is to conduct experiments with these other levels and populations. This is precisely what has happened with sentence-combining research, which has spawned numerous studies at other levels and with other student populations (see Daiker, Kerek, and Morenberg, 1979).

Several other kinds of threats can weaken external validity.

REACTIVITY ARRANGEMENTS

Reactivity Arrangements, sometimes called the Hawthorne effect, occur simply as a result of doing research on people. This threat is the inability to generalize research results to the real world of practice and decision making because of several factors: the artificiality of the experimental setting, the subjects' knowledge that they are participating in an experiment, attempts by the subjects to outguess the researchers, or whatever in the experimental situation is unrepresentative of real-world conditions. This threat has been shown, however, not to be a major one in the behavioral and social sciences, especially when research is done in school settings where assignment to classes, instructional "treatments," and criteria examinations are normal occurrences. Because all three studies were done in everyday classroom settings, it is unreasonable to assume that Reactivity was a threat to the external validity of their research.

INTERACTION THREATS TO EXTERNAL VALIDITY

Interaction threats also play a key role in a researcher's ability to generalize. These threats arise from combinations of threats to external and internal validity. If they exist, they jeopardize the ability to generalize. The most frequently listed interactions are Testing and Treatment, Selection and Treatment, Setting and Treatment, and History and Treatment. Because pretesting is not frequently used as an ordinary classroom practice, it may restrict generalizations in research studies that use it. If combined with other threats, it may cause unusual results over and beyond the simple effects of the threats alone.

In a true experiment like Halpern and Liggett's, there was no threat from an interaction between Treatment and Testing, because they did not pretest. A Selection and Treatment interaction was not a problem because the students were not selected, but were randomly allocated to the research groups. No unusual events occurred during the course of the research that influenced dictation skills. Setting and Treatment were not threats to external validity because the study was conducted in a university environment. In the studies of O'Hare and Hillocks et al., the threat of pretesting may have been avoided because their pretests were regular assignments.

Advantages and limitations of true experiments

The greatest advantage of the true experiment is its power to suggest cause-and-effect relationships because it can avoid many threats to internal validity. The possible exception is the threat of Mortality. In true experiments with no pretest, if there is considerable Mortality, it is difficult to determine what the scores or ratings would have been for those students who dropped out. But if there are differences among the students who dropped out, then the criteria measures will be biased. If pretests have been given, then a researcher can test for possible differences between the dropouts and those who remained in the experiment. A second exception occurs when the randomization process alerts students to the research condition that then may cause Reactivity or Social-psychological threats to internal validity.

True experiments are highly useful for composition research because they fit well into typical educational settings. The use of only a posttest saves the cost and time of developing pretest scores. The statistical analysis is also easier for a posttest-only true experiment. Its results can be interpreted more readily by those not trained in behavioral and social science research. The best interpretation of differences among the groups on the criteria is that these are cause-and-effect relationships between the treatment and the criteria.

Such conclusions are important for making academic decisions about pedagogy. As composition research develops, questions need to be raised about the ethics of using instructional methods that have not been empirically tested. Now, however, as new pedagogies based on new theories of writing are being introduced and tested, the old methods are also being tested. This has already happened with research on sentence-combining, which has called into question the value of grammar instruction. We are just beginning to find ways of testing instruction that aims to facilitate such acts as planning, audience-adapting, and revising.

One limitation of true experiments, however, is their focus on isolated variables and structured situations in order to determine possible cause-and-effect relationships. Because this precision turns attention away from the rich context, it is important for composition researchers to maintain a strong reciprocity between experimental and descriptive research.

Experimental and descriptive research

Maintaining a strong relationship between true experiments and descriptive research is important. The more researchers know from descriptive research about the nature of composing, the development of literacy, the influence of context on writing, and the nature of writing disabilities, the better will be the pedagogical methods that should be experimentally tested. Unlike researchers in other fields who do only qualitative basic research, most composition researchers must also carry on the responsibility of teaching composition. They cannot wait to conduct experiments until all variables have been fully defined and studied by descriptive research. They need to test continuously the efficacy of their methods by engaging in experimental research simultaneously with qualitative research. Both kinds of research can help to discover new variables.

Through the reciprocal use of both experimental and descriptive research, without glorifying either, researchers can attempt to avoid the criticism leveled by Emig (1982) and Mishler (1979) who argue that experimental research strips away the context of the natural environment and assumes that the important variables are known. True experiments should regularly be done in natural environments. Researchers must present enough "thick description" of their treatment conditions to increase theoretical understanding and to allow for replication. Experimental research therefore does not have to strip the situation of its rich features. Important new variables are as readily discovered in the context of true experiments as anywhere else. Moreover, because the cause-and-effect relationships among new variables are more evident in true experiments, they can provide a more conclusive substantiation of composition theories.

Summary

This chapter has presented the true experiment as a major type of experimental research available for composition studies. The features of a true experiment include randomization, treatments, hypotheses, criterion variables and analyses to determine statistical significance and effect size. Randomization helps to avoid the threats to internal and external validity. Below are some questions that a reader might ask when reading a report of a true experiment.

Checklist for reading "True Experiments"

1. What were the hypotheses? From what sources did they spring? Did they deal with significant problems in composition?
2. What were the treatment and control groups? Were they randomized? By classes or individuals?
3. What was the treatment? What were its sources?
4. What criterion variables were used?
5. What measurements were used to assess the criterion variables? Were they reliable and valid?
6. What were the statistical analyses? Did Type I or Type II errors occur? Was significance achieved? What were the effect sizes?
7. Did the study have internal and external validity?

References

Research Design and References

Asher, J.W. (1976). *Educational research and evaluation methods*. Boston: Little, Brown.

Bereiter, C., & Scardamalia, M. (1983). Levels of inquiry in writing research. In P. Mosenthal, L. Tamor, and S. Walmsley (Eds.). *Research on writing:* Principles and methods. New York: Longman.

Campbell, D.T., & Stanley, J.C. (1963). Experimental and quasi-experimental designs for research. In N.L. Gage (Ed.). *Handbook of research on teaching*. Chicago: Rand McNally, pp. 171–246.

Cohen, J.A. (1977). *Statistical power analysis for the behavioral sciences*. New York: Academic Press.

Cook, T.D., & Campbell, D.T. (1979). The conduct of randomized experiments. *Quasi-experimentation*. Boston: Houghton Mifflin.

Daly, J.A., & Hexamer, A. (1983). Statistical power in research in English education. *RTE, 17*(2), 157–164.

Edwards, A.E. (1985). *Experimental design in psychological research*, 5th edition. New York: Harper & Row.

Emig, J. (1982). Inquiry paradigms and writing. *CCC, 33,* 64–75.

Glass, G.V, McGaw, B., & Smith, M.L. (1981). Meta-analysis in social research. Beverly Hills, Calif.: Sage Publications.

Lana, R.E. (1959). A further investigation of the pretest treatment effect. *Journal of Applied Psychology, 43,* 421–422.

Mishler, E. (1979). Meaning in context. Is there any other kind? *Harvard Educational Review, 49,* 1–19.

Odell, L. (1981). Defining and assessing competence in writing. In C. Cooper (Ed.). *The nature and measurement of competency in English*. Urbana, Ill.: National Council of Teachers of English, pp. 95–138.

Quellmalz, E. (1981). Issues in designing instructional research: Examples from research on writing competence. Paper presented at the annual meeting of the American Psychological Association, Los Angeles.

Willson, V.L., & Putnam, R.R. (1982). A meta-analysis of pretest sensitization effects in experimental research. *American Educational Research Journal, 19*(2), 249–258.

Winer, B.J. (1971). *Statistical principles in experimental design*. New York: McGraw-Hill.

Witte, S., & Davis, A. (1982). The stability of T-unit length in the written discourse of college freshmen: A second study. *RTE, 16,* 71–84.

Young, R.E., & Koen, F. (1973). *The tagmemic discovery procedure: An evaluation of its uses in the teaching of rhetoric*. University of Michigan, Ann Arbor. National Endowment for the Humanities Grant No. EO 528-71-116.

True Experiments

Alverman, D.E., & Boothby, P.R. (1984). *Knowledge of text structure and its influence on a transfer task*. ERIC Document Reproduction Services No. ED 243 081.

Baker, C.A. (1984). Effects of comparison/contrast writing instruction on the reading comprehension of tenth grade students. *DAI, 45,* 09A.

Bateman, D.R., & Zidonis, F.J. (1966). *The effect of a study of transformational grammar on writing*. NCTE Research Report No. 6. Urbana, Ill.: National Council of Teachers of English.

Beach, R. (1979). The effects of between-draft teacher evaluation versus student self-evaluation on high school students' revising of rough drafts. *RTE, 13,* 111–120.

Boutte, M.A. (1979). The effects of cognitive behavior modification on enhancing creativity. *DAI, 40,* 11A.

Brossell, G. (1983). Rhetorical specification in essay examination topics. *College English, 45,* 165–173.

Buxton, E.W. (1958). An experiment to test the effects of writing frequency and guided practice upon students' skill in written expression. *DAI, 19,* 04.

Clifford, J. (1981). Composing in stages: The effects of a collaborative pedagogy. *RTE, 15,* 37–44.

Crowhurst, M. (1987). Cohesion in argument and narration at three grade levels. *RTE, 21,* 185–201.

Culp, M.M., & Spann, S. (1984). *The influence of writing on reading*. ERIC Document Reproduction Service No. ED 243 083.

Daiker, D., Kerek, A., & Morenberg, M. (Eds.). (1979). *Sentence-combining and the teaching of English*. Conway, Ark.: L&S Books.

Daly, J., & Dickson-Markman, F. (1982). Contrast effects in evaluation essays. *Journal of Educational Measurement, 19,* 309–316.

Day, M.M. (1983). Characteristics of the concept of audience in fifth grade writing. *DAI, 45,* 01A.

Dean, R.S. (1984). Cerebral laterality effects in the dual processing of prose. *Contemporary Educational Psaychology, 9*(4), 384–393.

Elley, W., Barham, J. Lamb, H. Wylie, M. (1976). The role of grammar in a secondary school English curriculum. *RTE, 10,* 5–21.

Faigley, L. (1979). The influence of generative rhetoric on the syntactic maturity and writing effectiveness of college freshmen. *RTE, 13,* 197–206.

Fitzgerald, J., & Teasley, A.B. (1983). *Effects of instruction in narrative structure on children's writing*. ERIC Document Reproduction Service No. ED 243 076.

Ghomi, C. (1984). Quality of prose recall as a function of cognitive style. *DAI, 45,* 05A.

Halpern, J., & Liggett, S. (1984). *Computers and composing: How the new technologies are changing writing*. Carbondale, Ill.: Southern Illinois Univ. Press.

Hawisher, G.E. (1987). The effects of word processing on the revision strategies of college freshman. *RTE, 21,* 121–144.

Hayes, I. (1984). An experimental study of sentence combining as a means of improving syntactic maturity, writing quality, and grammatical fluency in the compositions of remedial high school students. *DAI, 45,* 08A.

Hilgers, T. (1982). Training college students in the use of freewriting and problem-solving heuristics for rhetorical invention. *RTE, 14,* 293–307.

Hillocks, G. (1982). The interaction of instruction, teacher comment, and revision in teaching the composing process. *RTE, 16,* 261–278.

Hillocks, G., Kahn, E., & Johannessen, L. (1983). Teaching defining strategies as a mode of inquiry. *RTE, 17,* 275–284.

Holmes, J.G. (1984). An experimental study of the effects of independent writing time and exposure to a writing role model on selected kindergarten children. *DAI, 45,* 07A.

Hull, G. (1981). Effects of self-management strategies on journal writing by college freshmen. *RTE, 15,* 135–148.

Jordon, M.K. (1983). The effects of cooperative peer review on college students enrolled in required advanced technical writing courses. *DAI, 45,* 05A.

Kegley, P.H. (1984). The effect of mode of discourse on student writing performance. *DAI, 45,* 05A.

Kelley, K.R. (1984). The effect of writing instruction on reading comprehension and story writing ability. *DAI, 45,* 06A.

Kennedy, G.E. (1983). *The nature and quality of compensatory oral expression and its effects on writing in students of college composition*. ERIC Document Reproduction Service No. ED 240 597.

Koenig, J.L. (1984). Enhancement of middle school students' written production through the use of word processing. *DAI, 45,* 09A.

Kroll, B. (1978). Cognitive egocentrism and the problem of audience awareness in written discourse. *RTE, 12,* 269–281.

Land, R.E. (1984). Effect of varied teacher cues on higher and lower ability seventh and eleventh grade students' revision of their descriptive essays. *DAI, 45,* 05A.

Langer, J. (1984). Examining background knowledge and text comprehension. *Reading Research Quarterly, 19,* 468–481.

Lauer, J.M., Atwill, J., McCoy, N., & Rosenthal, A. (1986). *Writing as inquiry: Strategies and measures.* Paper presented at the CCCC meeting, Philadelphia.

McClaran, N.E. (1984). An examination of the effectiveness of the Marshall writing workshop approach in teaching writing. *DAI, 45,* 08A.

Mulder, J., Brown, C., & Holliday, W. (1982). Effects of sentence-combining practice and linguistic maturity levels in students. *Adult Education, 28,* 111–120.

Neville, D.D., & Searls, E.F. (1985). The effect of sentence-combining and kernel-identification training on the syntactic component of reading comprehension. *RTE, 19,*(1), 37–61.

O'Hare, F. (1973). *Sentence-combining: Improving student writing without formal grammar instruction.* Urbana, Ill.: National Council of Teachers of English.

Pelligrini, A.D. (1984). Identifying causal elements in the thematic-fantasy play production paradigm. *American Educational Research Journal, 21*(3), 691–701.

Piché, G., Rubin, D., & Turner, L.J. (1980). Training for referential communication accuracy in writing. *RTE, 14,* 309–318.

Piché, G., Rubin, D. Turner, L.J., & Michlin, M. (1978). Teachers' subjective evaluations of standard and black non-standard English compositions—a study of written language attitudes. *RTE, 12,* 107–118.

Raforth, B.A., & Rubin, D.L. (1984). The impact of content and mechanics on judgments of writing quality. *Written Communication, 1*(4), 446–458.

Reed, W.M. (1984). The effects of writing ability and mode of discourse on cognitive capacity engagement. *DAI, 45,* 07A.

Rubin, D., & Piché, G. (1984). The effects of available information on responses to school writing. *RTE, 18,* 8–26.

Quellmalz, E., Capell, F., & Chou, C. (1982). Effects of discourse and response mode on the measurement of writing competence. *Journal of Educational Measurement, 19,* 242–258.

Schiff, P. (1978). Problem solving and the composition model: Reorganization, manipulation, and analysis. *RTE, 12,* 203–210.

Smith, W.L. (1985). Some effects of varying the structure of a topic on college students' writing. *Written Communication, 2*(1), 73–89.

Stevens, R.J. (1983). Strategies for identifying the main idea of expository passages: An experimental study. *DAI, 45,* 01A.

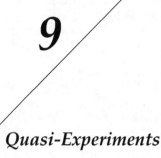

9

Quasi-Experiments

Quasi-experiments are a second general type of experimental research. These kinds of experiments are useful when researchers cannot randomize groups, for example, when classes must be kept intact. Much like true experiments, they have subjects, treatments, and criterion measurements. Quasi-experiments enable researchers to make cause-and-effect inferences about the effects of treatments on criterion variables. Three features distinguish the quasi-experiment from the true experiment.

1. There is *no* randomization of subjects to groups but rather the use of already established, intact groups.
2. The quasi-experiment *must* have at least one pretest or prior set of observations on the subjects in order to determine whether the groups are initially equal or unequal on specific variables tested in the pretest.
3. There must be research design hypotheses to account for ineffective treatments and threats to internal validity.

As indicated in the third feature above, investigators must give considerable attention to avoiding the threats to internal validity discussed in Chapter 8. In true experiments, most of these internal validity threats are of minor importance because randomization has equalized them in both the experimental and control groups. In the quasi-experiment, however, many of these threats may afflict the intact groups. Researchers must determine at the outset the possible effects on the criterion variables of these threats to validity in order to avoid confusing any apparent treatment effects with the effects of threats to internal validity.

Three quasi-experiments

Study 1   Roy Fox (1980). Treatment of writing apprehension
and its effects on composition. *RTE, 14,* 39–49.

Fox investigated the effect of two methods of teaching writing on three criterion variables: writing apprehension, overall writing quality, and length of writing. The control group received traditional instruction by means of exercises, lectures, discussions, and question-and-answer sessions, with emphasis on the modes of development and teacher evaluation. The experimental treatment was a workshop method using large-group interaction exercises, instructor-taught objectives for each essay, paired-student and small-group language problem-solving activities, free-writing, practice responses to writing, structured peer response, and two instructor–student conferences. The hypotheses were as follows:

1. Students in the experimental group would report significant reduction in writing apprehension as measured by pre-and post-Writing Apprehension Tests.
2. Students in the control group would retain their original levels of writing apprehension, as measured by the same tests.
3. Students (not just the writing-apprehensive students) in the experimental group would report significantly lower levels of writing apprehension at the end of the study than would all students in the control group.
4. Students ranked highest in writing apprehension at the beginning of the study in the experimental group would report significantly lower levels of writing apprehension at the end of the study than would students similarly ranked in the control group.
5. Students ranked highest in writing apprehension at the beginning would write posttest compositions that would be evaluated by two independent judges as significantly higher in quality than would those in the control group similarly ranked.
6. Students in the experimental groups would write better posttest compositions than the control group.
7. Students who ranked highest in writing apprehension at the beginning in the experimental group would write significantly longer posttest compositions than would those similarly ranked in the control group.

8. Students in the experimental group would write significantly longer compositions than the control group.

Fox used six intact classes (total $N = 106$) of freshmen male and female students in English composition at the University of Missouri— Columbia during the spring semester of 1978. All students had T scores of below 49 on the Missouri College English Placement Test. He excluded students who had previously failed Composition 1, who had excessive absences, or who had dropped out before the end of the semester. As necessary in a quasi-experiment, he administered as a pretest the Daly–Miller Writing Apprehension Test for which an analysis of variance revealed no significant difference between the treatment and control groups.

Three graduate instructors taught one experimental and one control section; they had no previous experience with the experimental treatment but did have with the control treatment. Each control group completed five outside-of-class essays along with in-class writing. The three experimental groups completed seven out-of-class essays. A tally of the words written in each group indicated them to be approximately equal. Both groups chose from the same pool of topics, the control group members choosing their own, and the experimental group choosing by consensus. Both groups had an equal emphasis on grades. As the hypotheses indicate, the criterion variables were level of writing apprehension, quality of posttest, and length of posttest. To measure these, the researcher used the Daly–Miller Writing Apprehension Test and a two-hour carefully controlled posttest writing sample. Two trained raters holistically scored the samples. The interrater reliability correlation was .92. For four of the eight hypotheses (1, 3, 4, and 8), the researcher found statistically significant differences between the experimental and control groups. Fox arrived at three main conclusions: (1) either method reduces writing apprehension; (2) the experimental method significantly reduced apprehension at a faster rate; and (3) the experimental treatment produced writing at least as proficient in quality as that of the control group.

Study 2 Marion Crowhurst and Gene Piché (1979). Audience and mode of discourse effects on syntactic complexity in writing at two grade levels. *RTE, 13*, 101–110.

Crowhurst and Piché did an experiment, part true and part quasi-experimental (they can be readily mixed), to examine the effect of intended audience and mode of discourse on the syntactic complexity of compositions written by sixth and tenth graders. The subjects were sixty boys and girls in each of the two grades. Even though the subjects could not be randomly assigned to a combination of the sixth and tenth grades (and clearly not to male and female), the grade–sex clusters were assigned randomly to the two audiences and three discourse modes. In other words, the students were assigned randomly within grade and gender groups to one of three mode conditions and two audiences. Thus the audience and mode comparisons had all the strengths of a true experiment. The comparisons between grade and sex stemmed from a quasi-experimental design. All students were pretested for syntactic complexity using Hunt's Aluminum Passage. Within grade levels, the T-unit lengths for the three groups were equal. The sixth-grade means within each modal group were 6.56, 6.50, and 6.53. The tenth-grade means were 10.15, 9.89, and 10.07. The pretest also showed no differences between boys and girls.

The treatment consisted of two conditions: audience (best friend and teacher) and three modes of discourse (narration, description, and argument). Each student wrote on each of three topics—canoe, classroom, and whale—in their assigned mode for two audiences. Pilot studies had shown that these topics produced writing samples of about 400 words. The series of six assignments also acted as pretests and posttests in that the pattern of changes from assignment to assignment between grades was observed over the various treatments: mode, audience, and topic. The criterion variables were mean number of words per T-unit, mean number of words per clause, and mean number of clauses per T-unit—all measures of syntactic complexity. Thus both the treatment and criterion were "multivariate"; more than a single variable for each was used. The methods of analysis and results will be discussed below.

Study 3 William Smith, and Warren Combs, (1980). The effects of overt and covert cues on written syntax. *RTE, 14,* 19–38.

In two experiments, Smith and Combs studied the effects of three instructional cues—overt, covert, and none—on the writing of ten intact composition classes. The first experiment studied the effect of uncued and overtly cued instructions on the rewriting of the Aluminum Passage by four classes. On the second day, two sections of students rewrote the controlled-stimulus passage (CSP) with overt cues and two without. After two days of regular instruction, one overtly cued section wrote with no cue and one with overt cues; one uncued section wrote with cues and one without. The overt cue consisted of an instruction indicating that the reader was an intelligent person influenced by long, complex sentences.

The second experiment studied the effect of covert cued instructions on the CSP and the free-writing of six sections of students. On the first day, two sections of students did uncued free-writing and the CSP and two did overtly cued free-writing and the CSP. After two days of covert cueing using a programmed sentence-combining text, one uncued section wrote without cues, one uncued section with overt cues, one overtly cued section with no cues, and one overtly cued section with overt cues. Their experimental design was as shown in Tables 9-1 and 9-2.

Table 9-1 Design for Experiment #1

Section	N	Day 2	Day 5
A	18	no cue	no cue
B	20	no cue	overt cue*
C	19	overt cue*	no cue
D	18	overt cue*	overt cue*

Source: Smith and Combs (1980), Figure 1. Copyright © 1980 by the National Council of Teachers of English. Reprinted by permission of the publisher.

*Students were told that the reader would be a highly intelligent person who is swayed by long, complex sentences.

Table 9-2 Design of Experiment #2

			Day	
Section	N	1–2*	3–4	5–6*
E	17	no cue	SC	no cue
F	20	no cue	SC	overt cue**
G	18	overt cue**	SC	no cue
H	18	overt cue**	SC	overt cue**
I	17		SC	no cue
J	15		SC	overt cue**

Source: Smith and Combs (1980), Figure 2. Copyright © 1980 by the National Council of Teachers of English. Reprinted by permission of the publisher.

*A free writing sample and rewrite of a controlled stimulus passage (CSP) were elicited.
**Students were told that the reader would be a highly intelligent person who is swayed by long, complex sentences.

Research design: types of quasi-experiments

Strong quasi-experiments

Quasi-experiments can be broadly classified into two categories: strong and weak, based largely on the equality or inequality of the groups as established by the pretest. Strong quasi-experiments mimic—in most cases rather well—the true experiment, because the pretest and observations show that the groups are equal on the variables measured. Their cause-and-effect logic is the same: equal groups acted on by equal treatments produce equal results on the criteria. Or, equal groups when acted on by unequal treatments produce inequality on at least one criterion variable. (In actuality, this is a research inference made about treatment effects, because the effects of a treatment or condition are actually unknown until criterion variables among several treatment groups are shown to be different or not different.)

In the strong quasi-experiment, the use of pretests and observations serves to establish the equality of intact groups at least on the variables represented by the pretests. In the true experiment, randomization allows the researcher to claim that *all* variables, known and unknown, upon which humans can differ will be expected to be equal in all treatment and control groups, both at the time of the randomization and for all future time unless there is an unequal intervention treatment. Further, randomization is expected to equalize all prior background variables including heredity and cultural environment. For any experiment, these statements of equalities are better stated as "not being different except as would be expected by chance"—with chance being defined in terms of statistical probability.

In the strong quasi-experiment, the researcher can consider equal only those explicit variables measured by the pretests. In practice, this determination is not of great concern because of the limited number of variables that correlate even moderately with the criterion variables. These criterion variables usually can be captured with a relatively small number of pretests or observations. In fact, the rule for choosing pretests and observations is to choose those that are most likely to correlate with the criterion variables. If on the pretest variables there are no differences among the groups, it is likely that there are no initial differences among these groups as well.

Crowhurst and Piché chose one writing sample, the Aluminum Passage, on which to compare the variable of syntactic complexity. On the variable, sex, the pretest showed no significant differences between the males and females on the criterion variable of syntactic complexity. Sex then became one variable aiding in defining a strong quasi-experiment. Fox also had a strong quasi-experiment because his pretest measures showed his groups to be equal. Part of Smith and Combs' study was a strong quasi-experiment. In Experiment 1, on question 1, the means of combined sections A and D were similar to the means of combined sections B and C on two criteria measures—words per T-unit (W/T) (12.61 and 12.16) and words per clause (W/C) (8.66 and 8.93). In Experiment 2, one pair of class sections, E and F, was essentially equivalent to the second pair, G and H: 14.65 and 14.96 and 17.26 and 17.15 on W/T; 8.11 and 8.29 and 8.89 and 8.92 on W/C.

Weak quasi-experiments

Weak quasi-experiments have initially unequal groups. On the pretesting, the intact groups are shown to differ on one or more variables. With initially unequal groups, the research inference logic changes: unequal groups acted upon by equal treatments will still be equally unequal on the criterion variables; or, unequal groups acted upon by unequal treatments will be either more unequal or less unequal on the criterion variables. One type of weak quasi-experiment has a particularly powerful logic. If Treatment Group B which is lower on the pretest than Group A, is markedly better on the criterion variable than Group A, then the treatment given to Group B had a powerful beneficial influence.

In the Crowhurst and Piché design, the variable, grade, defined a weak quasi-experiment because the pretest showed the sixth and tenth graders to have unequal syntactic complexity. Part of the Smith and Combs study was weak quasi-experimental because the class sections were probably not equal at the start of the research. They made comparisons by contrasting pairs of differences, before and after conditions, between groups of class sections. Tables 9-3 and 9-4 show how they pooled sections for different treatments. As shown in Table 9-3, the W/T criterion mean was 10.77 for no cue and 13.92 for overt cue; the W/C criterion mean was 8.17 for no cue and 9.45 for overt cues. In Table 9-4, the means were 14.81 and 17.21 and 8.21 and 8.91, respectively. Given the standard deviations of these pooled groups on the two criteria, it is reasonable to assume that they differed initially.

The initial scores on syntactic complexity—some under treatment conditions—served as the pretest measure of a quasi-experiment. Nowhere did the researchers declare that the intact classes were allocated to the treatment conditions at random. However, if they had simply

Table 9-3 Means and Standard Errors for Pooled Day 1–2
No Cue and Overt Cue CSP Writing (Experiments #1 and #2)

Section	N	W/T		W/C	
		Mean	SD	Mean	SD
Experiment #1 (pooled sections)					
No cue	38	10.77	2.22	8.17	1.60
Overt cue	37	13.92	3.16	9.45	2.01
Experiment #2 (pooled sections)					
No cue	37	11.63	2.19	7.71	1.03
Overt cue	36	14.48	2.70	9.80	1.98
All sections					
No cue	75	11.19	2.17	7.94	1.39
Overt cue	73	14.20	3.03	9.62	2.05

Source: Smith and Combs (1980), Table 2. Copyright © 1980 by the National Council of Teachers of English. Reprinted by permission of the publisher.

Table 9-4 Means and Standard Deviations for Pooled Day 1–2
No Cue and Overt Cue FREE Writing

Section	N	W/T		W/C	
		Mean	SD	Mean	SD
No cue	37	14.81	2.68	8.21	1.22
Overt cue	36	17.21	2.58	8.91	1.50

Source: Smith and Combs (1980), Table 5. Copyright © 1980 by the National Council of Teachers of English. Reprinted by permission of the publisher.

assigned the treatment conditions at random to the ten intact classes, the design would have been simpler and stronger. Then initial equality among the class sections on all variables—relevant and irrelevant—would have been assumed, causing all later differences among the criterion variables for the ten classes to be considered the result of the effects of the treatments. The degrees of freedom (see Appendix I) are less, however, when randomizing classes to treatments than when randomizing students to groups because the degrees of freedom are based on the number of class means. On the other hand, class means, with an average of 18 students in each class, can be very stable, less than one-fourth of the variability of individual students.

As it is, with initial differences among the groups on the measured variables, a reader must question whether other influential unmeasured variables also differed. Such inequality makes any eventual comparisons difficult to interpret. Measurement theory indicates that initially differing intact groups are expected to be somewhat more equal on any second or subsequent observations because of the Regression-Toward-the-Mean phenomenon. Observed gain scores determined simply by subtracting pretest scores from posttest scores are known to be biased, overestimates of true gain scores.

Other types of quasi-experiments

In addition to strong and weak quasi-experiments, several other types exist: the interrupted time series, the regression discontinuity, and the repeated-treatment designs.

INTERRUPTED TIME SERIES

The interrupted time series design can be used when no comparable groups are available. A way of conducting this kind of design is to administer a series of observations to establish a baseline of behavior and its natural variations under typical conditions without introducing any intervening experimental treatment. After four or five observations have taken place over an extended time period, a treatment is introduced, followed by a second series of observations, similar in number and timing to those prior to the treatment. If changes occur in the mean of the sets of observations from the pretests to the mean of the post-tests, then a researcher can conclude that the experimental condition caused the difference.

Several threats to internal validity attend this design. Maturation can be a problem if the design is used with children or adolescents. Testing is also a threat because of the introduction of so many tests prior to treatment. Mortality often occurs because of the time period needed to complete the research. Finally, the researcher may end up observing only a cyclical phenomenon, such as the rise of achievement during school sessions and a decline during the vacation (Cook and Campbell, 1979).

REPEATED-TREATMENT DESIGN

The Repeated-Treatment design is similar to Interrupted Time series. One or more observations are made prior to the introduction of a treatment. Then the treatment is introduced, followed by more observations, a withdrawal of treatment, a resumption of observations, and then a reapplication of the treatment, further observations, and so on. (This is the design often used by B.F. Skinner.) The principle is that if the treatment has an effect, the subsequent observations will change and when the treatment is withdrawn the observations will return to the original baseline. Of course if the results are permanent, the subsequent observations will stay changed (Vockell, 1983).

The Regression-Discontinuity design is useful in the study of the effects of remediation or scholarship awards in cases where insufficient funds exist to accommodate all students who might benefit from the aid of remediation. The researcher uses two regression equations (one for the aided group and one for the unaided group) to relate the measure by which students were selected with the measure of later accomplishment reflecting the goals of the remediation. If the remediation help is negligible, the two regression lines will be similar to an overall regression line fitted to the scores of all students. If the remediation has an effect, then these lines will differ (Asher, 1976).

Preexperiments and pseudoexperiments

Preexperiments and pseudoexperiments are sometimes discussed in the context of quasi-experiments. A *preexperiment* is generally an inadequate design from which to make cause-and-effect statements. In preexperiments, only one group is observed, treated, and then observed again. It is often faulty to attribute changes occurring between the pre- and post- observations to the effects of the treatment. A *pseudoexperiment* is also a faulty design in which an experimentally treated group is compared to a nontreated group with no attempt to show that either group was initially equal or unequal, or that the criterion differences between the groups were either smaller or greater than the initial differences between the groups.

We identify these types of experimental designs here because they appear in the literature of composition studies. With no evidence of initial equality or inequality, a researcher, even with elaborate statistical tests, cannot assume that groups that are statistically significantly different on criterion variables were also not that way at some prior time. The research logic is marred by the threat of Selection. Groups that initially differ are likely still to be different at some future time whether or not an intervening treatment has occurred. If, however, a preexperiment or pseudoexperiment is considered as an exploratory study, the researcher can identify emerging interrelationships among variables, which can then be later experimentally tested.

In two of Smith and Combs' class sections, I and J, no writing samples were collected in the first two class days. Without pretesting, this
part of the study was not a quasi-experiment but a pseudoexperiment.
Despite this, the researchers statistically related section I with pooled
section E and F and statistically compared section J's writings on the
fifth and sixth days with pooled sections G and H and with sections C
and D (p. 26). Interpreting such statistical analyses and comparisons is
difficult if not impossible.

Research design: procedures

Statistical analyses

Because quasi-experiments require pretests and posttests, the statistical
analyses must be repeated-measures analysis of variance, analysis of
covariance, gain scores, or correlated data t-tests for each criterion variable. Each observation or score, both pretest and posttest, contributes
to the counting of the total degrees of freedom (see Appendix I) in a
statistical analysis. It is a major error in analysis not to use repeated-
measures analysis of variance or covariance analysis for quasi-
experiments (and true experiments using pretests). A full explanation
of the methods by which researchers take into account all the condition
and treatment variables in a repeated-measures analysis of variance or
analysis of covariance, or by which the total degrees of freedom are
counted and allocated can be found in any advanced statistical analysis
text (Winer, 1971; Edwards, 1985; Lindquist, 1953).

MULTIVARIATE ANALYSES

As we have said, true and quasi-experiments can have two, three, four,
and even more treatments in one study. Each of these treatments can
have several conditions or levels. A researcher can design a true or
quasi-experiment with two treatments simultaneously, one with a two-
level treatment, and one with a three-level treatment. Such an experiment creates a 2×3 design, with a six-cell matrix. In a true experiment,
a researcher can allocate sixty subjects equally to all six cells and assess
the results on the criterion variable. Note that essentially two research
studies can be conducted simultaneously. Conducting two studies together is far more efficient than doing them separately. Moreover, a
researcher can investigate, at only minor cost, the interrelationships,
the interactions, of the two treatments.
Statistical tests have been developed to study individual differences

among three or more means in levels of a treatment. Some of these comparison tests in the analysis of variance are the Newman–Keuls, Tukey, Bonferroni, and adjusted *t*-tests. (See Appendix I).

If a researcher studies only one treatment with two levels, the sixty subjects are randomly allocated to the two levels. But this is less effective than the somewhat more complex design discussed above. There are some restrictions on the more complex types of experiments. When a researcher uses only one treatment with several levels in a true experiment, the number of subjects per cell does not have to be equal as long as there are at least two per cell. In an experiment with two or more treatments (each of which has at least two levels), the computations are simpler if there are an equal number of subjects in each cell. If they are unequal, the researcher usually requires a computer program with an unequal-cells analysis of variance (Winer, 1971).

ANALYSES FOR STUDY 1

In quasi-experiments if repeated-measures analysis of variance or analysis of covariance are not used, nonsignificant results will be suspect. This is because the repeated-measures analysis of variance and the analysis of covariance reduce the size of the error terms of the *F*-test (see Appendix I), making the *F* ratios larger and more likely to be statistically significant. (Significant results usually only become more significant.) Because Fox did not use the pretests and essay ratings in a repeated-measurement analysis of variance or analysis of covariance, all of his nonsignificant results may be suspect.

He also reported as statistically significant a result of .056 that was not significant at the .05 level that he had preselected for all analyses of data (p. 46). With a *p* level preset at .05, two better solutions would have been to report exact *p* levels in all cases or to have selected a *p* level of significance of .10 or even .20. Because Fox had no interactions between Instructor × Treatment, he was able to generalize his results.

ANALYSES FOR STUDY 2

Crowhurst and Piché conducted a multivariate experiment, going beyond a two-treatment design. They had five treatment conditions in one experiment: Grade (two levels: sixth and tenth), Sex (two levels: male and female), Discourse Mode (three levels: narration, description, and argument), Audience (two levels: best friend and teacher), and Topics (three levels: canoe, classroom, and whale). These were analyzed on each of three criterion variables: words per clause (W/CL), words per T-unit (W/TU), and clauses per T-unit (CL/TU). The re-searchers could therefore study combinations of five variable interrela-tionships at only a minor loss of statistical power, gaining more infor-mation more quickly about the effects of various treatments on the criterion variables than is possible conducting five separate studies. This multivariate experiment also better simulates the real world of instruc-tion where factors do indeed operate together.

The researchers made 4320 observations: 60 subjects (in each grade/sex cell of two grades and two sexes) times three discourse modes times two audiences times three topics times three criterion measures. To help clarify the analysis, we offer Figure 9-1, a diagram of their design.

When researchers combine a quasi and a true experiment, they must be careful about the application of the treatment variables and later about the analysis of variance. In Figure 9-1, there are *within-subjects*

			Best Friend			Teacher			← Audience
			CA	CL	W	CA	CL	W	← Topic
Gr 10	M60	N20							
		D20							
		A20							
	F60	N20							
		D20							
120		A20							
Gr 6	M60	N20							
		D20							
		A20							
	F60	N20							
		D20							
120		A20							

N = 240 subjects   1440

Three criteria (correlated)

W/CI   1
W/TU 1
CL/TU 1

     $3 \times 1440 = 4320$ observations

Figure 9-1 Diagram and analysis table of the Crowhurst and Piché five-factor, three-criterion research design. Topics abbreviations: CA, canoe; CL, classroom; W, whale. Modes abbreviations: N, narrative; D, description; A, argument.

variables (Audience and Topic, displayed horizontally) and *among-subjects* variables (Grade, Sex, and Mode, displayed vertically). It should be noted that the writings for Audience (best friend and teacher) and for Topic (canoe, classroom, and whale) could have been used as among-subjects variables on the left of the diagram; and the Mode variable (narration, description, argument) could have been used as within-subjects variables (as repeated measures) across the top of the diagram. (However, if any recombination of the three had been used, the analysis would be different.) The point is that researchers can apply more powerful statistical tests to the variables of the within-subjects' groupings and can detect differences more easily among these treatments' levels than for the among-subjects' grouping. Also note that variables such as Grade and Sex can only be studied as among-subjects' variables and cannot be put into the repeated-measurement, within-subjects groupings, because subjects cannot write several compositions as females and then as males. The same is true of grades.

When repeated measures are used on each subject on the same criterion variable, the analysis must take into account the correlations generated among the repeated measures. If researchers do not do this, they can draw erroneous conclusions by reporting nonsignificant statistical tests. Repeated-measurement designs, which always occur in quasi-experiments, require repeated-measures analyses (or covariance). Thus, any quasi-experiment that does not use repeated-measures analysis or covariance is open to question. Crowhurst and Piché are to be commended for using repeated-measures analyses correctly.

Interactions

Interactions, as we have said in Chapter 8, are the combined effects of two or more treatment or condition variables. A significant interaction says that something is happening among variables, causing either greater or lesser effects in combination than would be expected from each one alone. When there are no interactions, researchers can conclude that the effects operate in a similar manner across levels of conditions and treatments and thus can make generalizations about a main-effect treatment without regard to other treatment effects. The absence of interaction constitutes the essence of scientific generalization that tries to state principles free of specific conditions.

Crowhurst and Piché had quite a number of interactions. They had a significant first-order interaction between the levels of Mode and Audience on the words-per-clause criterion. The increase in clause length varied for Mode and Audience (1) in the narrative-mode level (7.55 for Best Friend and 7.66 for Teacher), (2) in the descriptive-mode level (8.71

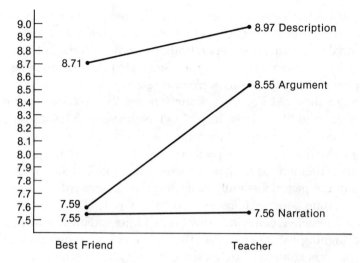

Figure 9-2   Significant interaction between mode and audience on clause length.

for Best Friend and 8.97 for Teacher), or (3) in the argument-mode level (7.59 for Best Friend and 8.55 for Teacher). Their results are illustrated in Figure 9–2.

Note the nonparallel lines in Figure 9-2 between the three pairs of points depicting the three mode levels over the two audience levels. It is this lack of parallelism that indicates an interaction. If there were no first-order interactions between Modes and Audience, the three lines would be parallel (or closer to parallel) and the researchers could say that any Mode had the same relative influence regardless of Audience, and vice versa. The authors indicated that "there were a number of significant two-, three-, and four-way interactions involving Topic," (p. 106) but suggested that because Topic was not a variable under examination in their study, they did not discuss these interactions.

Orders of Interaction

A *first-order* interaction is one between two treatment variables. A *second-order* interaction is one among three variables. A *third-order* interaction is one among four variables, and so forth. (Second-order interactions are difficult to interpret; third- and fourth-order interactions are even more difficult.)

In the Crowhurst and Piché study, there were 14 first-order interactions available for statistical testing, 11 second-order interactions, 5 third-order interactions, and 1 fourth-order interaction. Should a researcher want them, 31 statistical tests of hypotheses are available; in other words a great deal of information can be gained from even one criterion in one factorial experiment such as this one. Crowhurst and Piché were able to make considerable generalizations because they did not have many *significant* interactions. As they suggested, few of these second- and higher order interactions were of research interest.

Although they did not discuss the interactions of Topic with other variables, it would have been helpful to have their speculations on some likely causes of this unusual outcome, particularly when all other outcomes of the main variables seemed typical. Were there upper limits on the measurement used that restricted the range of criteria? Did one or more of the pictorial stimuli elicit greater or lesser syntactic complexity? Did certain combinations of Audience, Grade, Mode, or Sex produce the interactive behavior? This type of information would be helpful in designing future research. They were correct when they stated that Topic was controlled by crossing it with all the other variables, a statement true for all their variables and almost all experimental designs using analysis of variance.

ANALYSES FOR STUDY 3

To help the reader understand their treatment and data collection design, Smith and Combs provided figures and tables that we summarize in Figure 9-3.

One responsibility in the practice of empirical research is to report the procedures used in data collection and analysis in enough detail so that a knowledgeable person in the particular area can replicate the

CLASS DAYS

Class Section	1	2	3	4	5	6
Experiment 1						
A	CSP	CSP	NPA	NPA	CSP	CSP
B	CSP	CSP	NPA	NPA	CSP (cued)	CSP
C	CSP	CSP (cued)	NPA	NPA	CSP	CSP
D	CSP	CSP (cued)	NPA	NPA	CSP (cued)	CSP
Experiment 2						
E	CSP & free		SCP	SCP	CSP & free	
F	CSP & free		SCP	SCP	CSP & free (cued)	
G	CSP & free (cued)		SCP	SCP	CSP & free	
H	CSP & free (cued)		SCP	SCP	CSP & free (cued)	
I	Nothing		SCP	SCP	CSP & free	
J	Nothing		SCP	SCP	CSP & free (cued)	

Figure 9-3 Writing type, cueing, and activity type. All writing samples are evaluated on both W/T and W/C. Abbreviations: CSP, controlled-stimulus passage; NPA, nominal planned activities; SCP, sentence-combining practice.

research. Smith and Combs made clear their treatments and data collection methods but not their analytic procedures. They did have repeated measures over the six class days of data collection and treatments, and thus had correlated data. But it is difficult to know whether they used a repeated-measurement analysis for the analysis of variance. Not to have done so is usually an error, especially in cases where significance is not achieved. (If significance has been achieved, however, then the use of nonrepeated measures makes no difference because these analyses merely reduce the variance in the error terms, making the resulting F-ratios larger.)

Conclusions

Crowhurst and Piché discussed the variable, Grade, only in terms of total length of composition. Audience was significant for the criterion W/CL; W/TU "approached significance" (a phrase not compatible with good statistical interpretation when a significance level has been declared before the analysis). Sex was significant only on the criterion W/CL, while Topic was significant for the criteria W/TU and CL/TU.

Their three criterion variables—words per clause, words per T-unit, and clauses per T-unit—were correlated among themselves, suggesting that conceptually they were one criterion variable, represented by three operational definitions. In order to decide whether this was so, they could have intercorrelated these variables using factor analysis (see Appendix I) to determine the dimensionality of the criteria. Determining that each of the three independent criterion variables shows significant differences is far stronger than showing that three highly related criterion variables are significant. A multivariate criterion analysis of the analysis of variance is now being used regularly in this situation. Essentially the process entails (1) factor analysis of the several criterion variables, (2) the establishment of independent criterion variables, and (3) the use of each of these as an independent criterion variable in the multivariate treatment analysis of variance.

If Crowhurst and Piché had used multivariate criterion analysis instead of separate-criterion analysis of variance, significance levels reported for the resulting criterion variables would be independent of each other. As it was, because W/CL, W/TU, and CL/TU were evidently correlated, if a treatment variable was significant on one criterion variable, it would tend to be on the second and third criterion. For example the main effects of mode on each criterion variable—W/TU, W/CL, and CL/TU—were all significant beyond the $p < .001$ level.

Smith and Combs determined results by comparing class sections on the initial writing samples under the differing cueing conditions and by comparing gains from the pretest to posttest both from the initial day's writing samples to the final day's samples, and then across the classes' gains under the no cue, overt cue, and covert cue conditions. This was done with both the CSP and the free-writing samples using both the W/T and W/C criteria. The study's pretest comparisons, necessary for the research to be quasi-experimental, considerably strengthened the cause-and-effect statements that were made concerning cueing, rewriting, repeated writing, and the comparative gains among the ten sections. The researchers concluded that "repeated writing does not have an effect on rewriting the CSP, even in combination with a covert cue" (p. 29). But they represented evidence to suggest that at the beginning, the sections may not have been equal. Because there were only a few premeasures in the control classes, this makes it difficult to interpret the results of comparisons of among-class sections.

Reliability and validity

Reliability

As was indicated in Chapter 7, in all research studies it is imperative to know the psychometric characteristics of all variables: means, standard deviations, and internal-consistency reliabilities (Cronbach alphas or Kuder–Richardsons 20s). Among these reliabilities are interrater reliability, the reliability of various rater scoring methods, and the stability reliabilities of variables. Researchers and readers need to know these reliabilities in order to interpret correlations among variables. As indicated in prior chapters, correlations among variables cannot be high if these variables have low reliabilities. If the reliabilities of variables differ, researchers cannot interpret the same observed correlation between pairs of variables being of equal strength. Highly precise measurement of variables (indicated by the internal-consistency reliability coefficient) is impossible except for very long measures or very large numbers of observers. Thus, observed perfect relationships between variables are impossible. In this context, the reliabilities of the three studies are discussed below.

Fox used as one criterion variable, the Writing Apprehension Test, which has a well-established and repeated internal-consistency reliability of .94 via Cronbach's alpha. To establish the reliability of his other criterion, raters' evaluations of the final essay, Fox trained his raters carefully and achieved a reliability coefficient of .92 between their adjusted score ratings, certainly sufficient to be definitive in further analyses. His word count per essay was also highly reliable. He provided detailed descriptions of his treatment and control conditions to help others replicate them. Although a coefficient of reliability for these treatments generally cannot be calculated as such, a researcher can consider the treatment's reliability in terms of replicability.

Crowhurst and Piché had commendable interrater reliability. For scoring words per T-unit, words per clause, and clauses per T-unit, they reported reliabilities as .96–.99. For 5 to 10% of the writing samples, the reliability was .94 to .98. They did not, however, report any internal-consistency reliabilities for W/TU, W/CL, or CL/TU for these students' ranges of production or for the lengths of the measures used.

Smith and Combs, reporting correlations of students' performance on the CSP and Free-Writing (Table 8 in Smith and Combs, 1980), showed 20 observed correlations running from .69 to .96. Because both of these measures were short, their internal consistency reliabilities could not have been exceptionally high. This fact, coupled with some necessary interrater unreliability and some fluctuation of the subjects over days, suggests that with a median correlation value of .80, the CSP and Free-Writing samples were reflecting only one construct.

Internal validity

In quasi-experiments, the threats to internal validity, discussed in Chapter 8, can become serious problems because they have not been controlled by randomization.

Fox used an analysis of variance to establish the initial equality of the classes on the Writing Apprehension Test. As he reports, however, "it was necessary to eliminate some subjects high in writing apprehension" (p. 45), and those who had previously taken the course. This selective elimination of subjects on the basis of high (or low) test scores can cause problems of Regression-Toward-the-Mean. If the degree of anxiety and number of students was about the same in all classes, the effect would balance out, but this does not happen. It was not necessary for him to drop high-anxiety students to "achieve . . . initial relative equality across conditions" (p. 45). In fact, it is better to use unequal groups in a weak quasi-experiment design, because eliminations can reduce the generalizability of the results and because of problems with Regression-Toward-the-Mean. Finally, although he cast his research into a strong quasi-experimental design as the result of the eliminations through the pretest, he weakened the study because he only analyzed the posttest data. Hence he derived no analytic benefit from the pretest scores.

He also lost students who had excessive absences, dropped the course, or dropped out of school. He did not report the numbers of students in these categories or whether these dropouts differentially occurred. If they had been caused by the treatment, this experimental mortality would have influenced the results, which were based solely on those who remained fully involved to the end of the semester.

The Crowhurst and Piché study had negligible threats to internal validity. They had no problems with History, Maturation, Selection, or Regression-Toward-the-Mean. Their training of scorers and well-defined scoring methods eliminated the threat of Instrumentation. Because their study involved a routine educational activity, Reactivity was not a threat. Finally, the emotional threats of Diffusion, Rivalry, or Resentful Demoralization seem unlikely, although it would be useful to know why some 60 students wrote incomprehensible, nonsense, or obscene compositions and why there were so many higher-order significant interactions with Topic.

The study did have two possible threats, however. Mortality was a concern in that 359 students started the study and only 240 were in the final sample. About 60 of the students were "randomly discarded" (p. 102), so Mortality was not a threat for them. For about 60 others, however, some of whose work could not be used, it would be helpful to know that they were reasonably distributed among Modes, Topics, Audiences, Grades, and Sexes. Without this knowledge, the reader must consider the production of incomprehensible, nonsensical, and obscene compositions as a criterion variable in order to determine if one of the treatments caused this to happen. Instability probably did not occur. Given the probable reliability of the measures, the 240 subjects, the care to avoid statistical "fishing" for results, and the reasonable meeting of the assumptions of the statistical analysis, it was unlikely that a difference of any major magnitude among the levels of main variables went undetected or that nothing more than chance significance was found.

In the Smith and Combs study, the threats to internal validity made hazardous their numerous comparisons between the T/U and W/C counts for their college freshmen and the counts for subjects of several other researchers—Hunt's twelfth graders and skilled adults on the CSP (1970), Morenberg, Daiker, and Kerek's gain scores (1978), and Swan's (1977) results. The researchers did not control several threats to internal validity. To counteract the threat of Testing (the possible effect of the several compositions written in sequence), they merely concluded that the repeated trial did not affect the syntactic-complexity variable. They did not compensate well for the threat of Regression-Toward-the-Mean due to the initially unequal groups because they apparently did not use analysis of covariance or other regression methods to adjust the raw gain scores, which they directly compared with two of Hunt's studies and with those of Swan and Morenberg et al.

These threats constitute some of the main disadvantages of quasi-experiments. Other disadvantages as well as advantages will be discussed below.

Advantages and disadvantages of quasi-experiments

Because quasi-experiments can produce rather strong cause-and-effect statements between the treatment variables and criterion variables, they offer valuable additions to knowledge. A researcher cannot always conduct a true experiment, because many important variables cannot be randomly assigned to people: sex, socioeconomic status, intelligence, writing ability, ethnic background, age, childhood experiences, and so forth. In small schools and in classes where enrollments are not large, it also may be difficult to assign students randomly to sections of classes. Quasi-experiments are thus necessary in research on composition, although they may not produce as high a quality of evidence about cause-and-effect results as do true experiments. In quasi-experiments, the researcher must specifically identify the threats to internal validity which would possibly increase, decrease, or change the value of different criterion variables.

In weak quasi-experiments, in which the research inference logic further depends on the retention or increase of initial differences after the treatment, conclusions must be tentative. Further, the means of groups that differ initially (and that therefore may differ from the population means) will be expected to be less different on any subsequent measurement simply because of Regression-Toward-the-Mean. Thus initially high groups will be expected to decline after the treatment if that treatment has no effect. Similarly, initially low groups will regress up toward the mean.

Quasi-experiments are both useful and necessary. To obtain a better picture of cause-and-effect results in a given field, researchers can combine the results of different quasi-experiments, often balancing out major problems of interpretation that arise from the variations and types of settings of quasi-experiments. Combining the results of quasi-experiments and true experiments formally and statistically is called a meta-analysis, which estimates the effect sizes of the treatment effects and conditions in which the research was done (see Chapter 10). If investigators still have concerns about the possible effects of certain aspects of a series of quasi-experiments, they can rate the strength of the flaws of these features and then correlate these ratings with the effect sizes in each of the studies. If these correlations are essentially zero, researchers can conclude that the seeming flaws in the research design are of minimal importance or indeed have been balanced out. If, on the other hand, the correlations are not zero, they can identify the set of studies as flawed and exclude them from their theories of composition behavior.

The perfect design for any study either does not exist or will probably not be recognized until long after the data are collected. If researchers knew enough about the nature of a phenomenon to structure a perfect research design, the problem likely would no longer be of interest. A reasonable goal is to select the design that seems most appropriate at the outset and to conduct the research accurately within that design.

Summary

The quasi-experiment is a research design that uses intact groups, treatment and control, to make cause-and-effect statements. At the beginning, researchers give pretests to determine the equality of the groups and pose hypotheses to account for threats to internal validity. Two kinds of quasi-experiments exist, depending on the equality or inequality of the groups—strong or weak. With pretests and posttests, researchers must use repeated-measures analyses or analyses of covariance, making strong efforts to control the threats to internal validity.

Checklist for reading "Quasi-Experiments"

1. Did the study use intact, nonrandomized groups?
2. Were pretests given? What were they?
3. Was the experiment a strong or weak quasi-experiment?
4. What hypotheses were posed to account for possible threats to validity?

5. What were the treatments or conditions imposed? Were they multivariate?
6. What were the criterion variables?
7. What specific posttests or observations were given or made to measure the criterion variables?
8. What were the reliability and validity of all the variables in the study?
9. How did the study account for threats to internal and external validity?

References

Research Design

Asher, J.W. (1976). *Educational research and evaluation methods.* Boston: Little, Brown.
Campbell, D.T., & Stanley, J.C. (1963). Experimental and quasi-experimental designs for research. In N.L. Gage (Ed.). *Handbook of research on teaching.* Chicago: Rand McNally, pp. 171–246.
Cook, T.D., & Campbell, D.T. (1979). *Quasi-experimentation: Design and analysis issues for field settings.* Boston: Houghton Mifflin.
Edwards, A.L. (1970). *The measurement of personality traits by scales and inventories.* New York: Holt.
Edwards, A.L. (1985). *Experimental design in psychological research,* 5th edition. New York: Harper & Row.
Kenny, D.A. (1975). Quasi-experimental approach to assessing treatment effects in the nonequivalent control group design. *Psychological Bulletin, 82,* 345–362.
Lindquist, E.F. (1953). *Design and analysis of experiments in psychology and education.* Boston: Houghton Mifflin.
Stanley, J.C. (1966). A common class of pseudo-experiments. *American Educational Research Journal, 3,* 79–80.
Vockell, E.L. (1983). *Educational Research.* New York: Macmillan.
Winer, B.J. (1971). *Statistical principles in experimental design,* 2nd edition. New York: McGraw-Hill.

Quasi-Experiments

Austin, D.E. (1984). Reading to write: The effect of the analysis of essays on writing skills in college composition classes. *DAI, 45,* 03A.
Bair, M.R. (1984). A self-generated writing program and its effects on the writing and reading growth in kindergarten children. *DAI, 45,* 02A.
Bator, P. (1980). The impact of cognitive development upon audience awareness in the writing process. *DAI, 41,* 02A.
Beachem, M.T. (1984). An investigation of two writing process interventions on the rhetorical effectiveness of sixth-grade writers. *DAI, 45,* 08A.
Benson, N. (1979). The effects of peer feedback during the writing process on writing performance, revision behavior, and attitude toward writing. *DAI, 40,* 07A.
Burns, H. (1979). Stimulating rhetorical invention in English composition through computer-assisted instruction. *DAI, 40,* 07A.
Carroll, J. (1984). Process into product: Teacher awareness of how the writing process

affects students' written products. In R. Beach & L. Bridwell. (Eds.). *New directions in composition research*. New York: Guilford Press.

Combs, W. (1976). Further effects of sentence-combining practice on writing ability. *RTE, 10,* 137–149.

Combs, W. (1977). Sentence combining practice: Do gains in judgments of writing "quality" persist? *Journal of Educational Research, 70,* 318–321.

Crowhurst, M., & Piché, G. (1979). Audience and mode of discourse effects on syntactic complexity in writing at two grade levels. *RTE, 13,* 101–110.

Daiker, D., Kerek, A., & Morenberg, M. (1978). Sentence combining and syntactic maturity in Freshman English. *CCC, 29,* 36–41.

Delaney, M.C. (1980). A comparison of a student-centered free writing program with a teacher-centered rhetorical approach to teaching college composition. *DAI, 41,* 05A.

Dixon, D. (1984). The effect of poetry on figurative language usage in children's descriptive prose writing. *DAI, 45,* 08A.

Donlon, D. (1976). The effect of four types of music on spontaneous writing of high school students. *RTE, 10,* 116–126.

Dutch, W. (1980). A comparison of the use of student-generated heuristics with the use of Larson-generated heuristics in the college classroom. *DAI, 40,* 12A.

Fabien, M.G. (1984). Using a learning styles approach to teaching composition. *DAI, 45,* 07A.

Finnemore, S., Breunig, D.W., & Taylor, R.G. (1980). Effect of intensive instruction on ten measures of performance. *Reading Improvement, 17,* 158–162.

Fox, R. (1980). Treatment of writing apprehension and its effects on composition. *RTE, 14,* 39–49.

Graham, M.S. (1983). The effect of teacher feedback on the reduction of usage errors in junior college freshmen's writing. *DAI, 45,* 04A.

Hake, R., & Williams, J. (1979). Sentence expanding: Not can, or how, but when. In D. Daiker, A. Kerek, & M. Morenberg (Eds.). *Sentence-combining and the teaching of English.* Conway, Ark.: L&S Books.

Hanrahan, C.M. (1984). A comparison of two approaches to using writing across the curriculum. *DAI, 45,* 08A.

Hart, M.B. (1979). An experimental study in teaching business communications using two different approaches: Theory and application vs. writing. *Journal of Business Communication, 17,* 13–25.

Hayes, B.L. (1984). The effects of implementing process writing into a seventh grade English curriculum. *DAI, 45,* 09A.

Hillocks, G. (1979). The effects of observational activities on student writing. *RTE, 13,* 23–36.

Hofman, R. (1979). The relationship of reading comprehension to syntactic maturity and writing effectiveness. In D. Daiker, A. Kerek, & M. Morenberg (Eds.). *Sentence-combining and the teaching of English.* Conway, Ark.: L&S Books.

Kean, D.K. (1984). Persuasive writing: Role of writers' verbal ability and writing anxiety when working under time constraints. *DAI, 45,* 09A.

Lewis, R.M. (1984). Teaching English composition to developmental students at the college level: A free writing/language study approach versus a structured writing/language study. *DAI, 45,* 09A.

Markman, M.C. (1983). Teacher–student dialogue writing in a college composition course: Effects upon writing performance and attitudes. *DAI, 45,* 06A.

May, B.A. (1984). Effective instruction for teaching basic writing skills with computer-assisted language learning in an English as a second language program. *DAI, 45,* 12A.

Meiser, M.J. (1984). A cognitive-process approach to college composition: A comparative study of unskilled writers. *DAI, 45,* 07A.

Morenberg, M., Daiker, D., & Kerek, A. (1978). Sentence-combining at the college level: An experimental study. *RTE, 12,* 245–256.

Mosenthal, J.H. (1984). Instruction in the interpretation of a writer's argument: A training study. *DAI, 45,* 11A.

Mosenthal, P. (1984). The effect of classroom ideology on children's production of narrative text. *American Educational Research Journal, 21,* 679–680.

Nugent, S. (1981). A comparative analysis of two methods of invention. *DAI, 41,* 09A.

O'Donnell, A.M., Dansereau, D.F., Rocklin, T.R., Larson, C.O., Hythecker, V.I., Young, M.D., & Lambiotte, J.G. (1987). Effects of cooperative and individual rewriting on an instruction writing task. *Written Communication, 9,* 90–99.

Pederson, E. (1978). Improving syntactic and semantic fluency in the writing of language arts students through extended practice in sentence-combining. *DAI, 38,* 10A.

Piper, K.L. (1984). An investigation of the effects of sentence-combining and story expansion delivered by word-processing microcomputers on the writing ability of sixth graders. *DAI, 45,* 09A.

Potkewitz, L. (1984). The effect of writing instruction on the written language proficiency of fifth- and sixth-grade pupils in remedial reading. *DAI, 45,* 08A.

Powell, J.E. (1984). The effects of sentence-combining on the writing of basic writers in the community college. *DAI, 45,* 08A.

Prentice, W. (1980). The effects of intended audience and feedback on the writings of middle grade pupils. *DAI, 41,* 03A.

Sanders, S., & Littlefield, J. (1975). Perhaps test essays can reflect significant improvement in freshman composition: Report on a successful attempt. *DAI, 45,* 9, 145–153.

Smith, W., & Combs, W. (1980). The effects of overt and covert cues on written syntax. *RTE, 14,* 19–38.

Smith, W., & Hull, G. (1983). Direct and indirect measures for large-scale evaluation of writing. *RTE, 17,* 285–289.

Stewart, M. (1978). Freshman sentence-combining: A Canadian project. *RTE, 12,* 257–268.

Swann, M.B. (1977). The effects of instruction in transformational sentence-combining on the syntactic complexity and quality of college level writing. Unpublished doctoral dissertation, Boston University.

Thompson, R. (1981). Peer grading: Some promising advantages for composition research and the classroom. *RTE, 15,* 172–174.

Tierney, R., & Mosenthal, J. (1983). Cohesion and textual coherence. *RTE, 17,* 215–230.

Trageser, S. (1979). The student in the role of spectator-observer-commentator: The relationship between visual perception and specificity in writing *DAI, 40,* 05A.

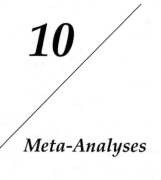

10

Meta-Analyses

After a series of true and quasi-experiments have been conducted on the effects of pedagogical methods or instructional environments, it is important to determine to what extent their conclusions are related. Meta-analysis, a major step forward in the methodology of the social and behavioral sciences, is a systematic and replicable way of summarizing the overall results of a particular body of research literature. In a sense it is now the ultimate way for researchers to generalize cause-and-effect statements from true and quasi-experiments, thereby establishing external validity. Because meta-analysis can enhance the generalizability of research conclusions, it can markedly develop theory about the behavior under study. Meta-analysis can also determine how conditions under which studies have been conducted have influenced their conclusions. We will use as an example of this design the study of Hillocks (1984), a meta-analysis in composition research.

One meta-analysis

Study 1   George Hillocks, (1984). What works in teaching composition: A meta-analysis of experimental treatment studies. *American Journal of Education, 93,*(1), 133–170.

Hillocks examined true and quasi-experimental research on three dimensions of composition.

1. Modes of instruction—presentational, natural process, individual, and environmental
2. Focus of instruction—grammar, models, sentence-combining, scales, inquiry, and free-writing
3. Duration of instruction

He selected 60 true and quasi-experimental studies with some 75 experimental versus control effect sizes. The studies had an average effect size result (in z-score terms) of +.28, which represents rather effective instructional methods. His design and results will be discussed throughout the chapter.

Ways of integrating studies

Combined significances

Until the development of meta-analysis, empirical researchers had no easy way to analyze rigorously the effectiveness of combined results of many studies in terms of the strength and consistency of their results. One attempt in the past was to combine the significance levels from independent studies. The p values of the statistical results of several studies were combined, but all this told the reviewing researcher was the overall probability of the set of results. Using this method, practically all sets of studies that yielded results in the same direction were statistically significant when combined

Box score method

Another quite common method has been to combine studies on the basis of judgments of experts. If they found a half-dozen or more studies on the same behavior, they compared the number of significant versus nonsignificant differences and drew conclusions. This method, sometimes called the "box score method," declared the results having the most significance the winner. The major problem with this method has been its inherent bias. Because the statistical results of behavioral and social science research as usually performed tend towards nonsignificant results, the box score method will lead generally to conclusions that are the reverse of what is true if indeed a treatment causes a difference on a criterion variable. The unreliability of the measurement systems and the difficulty of getting large numbers of subjects regularly mask true differences, even those of major magnitude. These problems cause Type II errors.

Meta-analysis

Meta-analysis is the best method so far devised to integrate studies. It addresses two major problems faced by the behavioral and social sciences: (1) the inability to detect differences in the usual samples, and (2) the finding of significance for almost all variables in very large samples. To assess the magnitude of the results of several studies, Glass (1976) developed effect sizes for multiple-group analysis-of-variance statistics and various kinds of correlational analyses to be used with true or quasi-experimental designs. This method, called meta-analysis, was explained in detail in its first major application (Smith, Glass, and Miller, 1980, pp. 213–217). Because the design is a relatively new methodology, developments, studies, and elaborations are published regularly. (See Hedges 1986; Hedges and Olkin, 1985; Glass, McGaw, and Smith, 1981).

Effect sizes are comparisons of the differences between groups with the natural individual differences of the subjects within groups. Effect size, a z score, is an equal-interval metric on a scale that goes from about −3.00 to 0 to +3.00. A ±3.00 effect size indicates an enormous difference between the groups, while a zero effect size indicates no difference between the groups. An effect size of about .20–.50 is considered important; anything above .50 is a major difference. Hillocks found strong effect sizes for the environmental mode of instruction (effect size .44) and for several focuses of instruction: inquiry (effect size .57), scales (effect size .36), sentence-combining (effect size .35), and the use of models (effect size .217).

Research design: meta-analyses

Study selection and criteria

In a meta-analysis, researchers first select the studies to be analyzed by an exhaustive search of the literature. To assure the similarity of the studies, researchers establish criteria by which the studies are chosen. A reasonable meta-analysis can be conducted on 10 or even 5 studies. Hillocks reviewed about 500 experimental studies, published between 1963 and 1982, and selected 60 studies, using the following criteria.

1. The study had to have a treatment over a period of time leading to a measured criterion variable following the treatment.
2. The study had to make use of a scale of writing quality applied to samples of writing (studies using standardized tests or matched pairs were excluded).
3. The study had to exercise minimal control for teacher bias.
4. The study had to control for differences among groups, i.e., either randomization, matching, or pretests.
5. Scoring had to be reliable and valid.

Variables

THE CRITERION VARIABLE

In meta-analysis, each research study under examination becomes *a sampling unit*. The effect size in each study becomes *the criterion variable*. In Hillocks' study, the 60 true and quasi-experiments became the sampling units. He calculated the effect size by dividing the difference between posttest scores, adjusted for the difference between pretest scores, by the pooled standard deviation of posttest scores for all groups in the study.

TREATMENT OR CONDITION

Selected conditions and treatments of each study become the constructs or variables whose influence the researcher wants to determine. The selection of these variables depends on the insight, creativity, and theoretical position of the researcher. Well-chosen variables can test and further develop theory in a field and can document gaps in the research literature. Hillocks identified as important the following treatment or condition variables: mode of instruction (presentational, natural process, and environmental), focus of instruction (grammar, models, sentence-combining, scales, inquiry, and free writing), and duration of instruction.

DATA MATRIX

After researchers have selected the variables and determined their homogeneity, they can then code and score these variables, entering these scores into a rectangular data matrix. The treatment or condition variables are put across the top along with the effect size. Each research study is put down the left side (see Figure 10-1).

	Effect size	Focus	Mode	Duration
Study 1				
Study 2				
Study 3				
Study 4				
$N=60$				

Figure 10-1   Data matrix for effect size.

Homogeneity

When these treatment and condition variables have been selected and entered into the data matrix, it is important to know whether their effects are homogeneous or similar. Hillocks, for example, wanted to determine whether all the treatments he had included under inquiry were indeed similar enough to belong there. To test for homogeneity, Hillocks used the "H"-test developed by Hedges (1981). The logic behind this test says that if any construct or treatment category such as inquiry is theoretically sound, then its embodiment in each study should produce similar effects on the criterion variable as measured by effect sizes. If treatment effects are similar, then they will be statistically homogeneous. The statistical test of homogeneity (H) will show no significant difference among them. Hillocks used this H-test to examine each of his treatment or condition variables. He found that the Natural Process Mode had significant variability or lack of homogeneity. He reduced this variability by removing the three studies that had the largest positive and negative effect sizes, noting that this could be done without appreciably affecting the overall average effect size for that treatment construct. Further, he explained that this removal was theoretically sound because the largest positive effect size in one study was due to a loss of effect size in a control group for which every mechanical and structural error in the compositions was marked and in which the treatment group made no gains. In a second study, the control group also lost on effect size, and in the third study, the treatment group had an effect size gain that was relatively large. Hillocks did not speculate about these last two major effect sizes because the studies' reports did not provide enough detailed information about the treat-

Table 10-1   Mode of Instruction: Summary of Experimental/Control Effect Size Statistics

	Mean Effect	SD	95% Confidence Interval		H	df	Maximum H for Nonsignificance at $p < .01$
			Lower	Upper			
All meta-analysis treatments ($N = 73$)	.28	.018	.24	.32	411.08	72	102.60
Treatments (four outliers removed ($N = 69$)	.24	.019	.20	.27	169.28	68	98.00
Treatments included in mode of instruction analysis ($N = 29$)	.24	.025	.19	.29	73.83	28	48.27
Natural process ($N = 9$)	.19	.037	.11	.26	23.15	8	20.08
Environmental ($N = 10$)	.44	.050	.34	.53	12.83	9	21.66
Presentational ($N = 4$)	.02	.114	−.20	.24	.92	3	11.33
Individualized ($N = 6$)	.17	.064	.06	.28	14.68	5	15.08
Treatments categorized by mode of instruction ($N = 29$)	. . .	. . .	. . .	. . .	51.58	25	44.3

Source: Hillocks (1984), Table 3. © 1984 by The University of Chicago Press. All rights reserved.

Table 10-2   Focus of Instruction: Summary of Experimental/Control Effect Size Statistics

Focus	Mean Effect	SD	95% Confidence Interval		H	df	Maximum H for Nonsignificance at $p < .01$
			Lower	Upper			
Treatments included in focus of instruction analysis ($N = 39$)	.26	.023	.21	.30	84.48	38	62.4
Grammar ($N = 5$)	−.29	.059	−.40	−.17	8.85	4	13.27
Sentence-combining ($N = 5$)	.35	.083	.19	.51	1.89	4	13.27
Models ($N = 7$)	.22	.057	.11	.33	5.31	6	16.80
Scales ($N = 6$)	.36	.078	.21	.51	6.89	5	15.08
Free writing ($N = 10$)	.16	.035	.09	.23	27.25	9	21.66
Inquiry ($N = 6$)	.56	.076	.41	.71	8.73	5	15.08
Treatments categorized by focus of instruction ($N = 39$)	. . .	. . .	. . .	. . .	58.92	33	54.70
With two outliers removed ($N = 37$)	. . .	. . .	. . .	. . .	49.90	31	52.15

Source: Hillocks (1984), Table 6. © 1984 by The University of Chicago Press. All rights reserved.

ments. He did, however, note that the homogeneity of Mode of Instruction as an entire category was not significant, a fact that resulted in similar effects in each mode of instruction. See Tables 10-1 and 10-2 for homogeneity statistics.

Statistical analyses: Experimental studies

CORRELATION

The original meta-analysis, developed by Glass (1978), used correlation (Smith, Glass, and Miller, 1980). This classic approach to meta-analysis involved an intercorrelation of the effect size criterion variable with all the conditions under which each study was done and published: quality of research design, format of publishing (thesis, journal, or book), duration of treatment, skill of instructor, type of raters, levels and quality of students, and so forth. Hillocks used correlation on the interval variable, Duration, to relate it with effect size. He got a correlation of $-.02$, which was of negligible importance given the range of duration. (Note that the averages of four categories of Duration—less than 13 weeks, more than 12 weeks, less than 17 weeks, or more than 16 weeks— showed no differences as would be expected with the near-zero correlation among the studies.)

ANALYSIS OF VARIANCE

Hedges and Olkin (1985) put the meta-analysis on a firmer statistical basis and expanded the approaches that researchers could take. They developed an approach using analysis of variance in which a researcher calculates the averages of the effect sizes of the treatment variables. Hedges also investigated the influence of the unreliability of criterion variables on the calculation of effect size (1981). (See also Hedges, 1986.) In composition studies that use criteria with lower reliabilities, ranging from .70 to .80, it may be appropriate to use Hedges' adjustments. Hillocks used Hedges' procedures to organize his selected studies into the four modes of instruction and six focuses of instruction mentioned above. He calculated the average effect sizes for these modes (see Figure 10-2) and focuses (see Figure 10-3). He then tested the several averages statistically for overall differences among categories and, for those that were significant, he tested for differences between any two categories and groupings of categories that seemed theoretically compatible.

For mode of instruction, his calculations revealed that presentational mode (lecturing) was of negligible effectiveness (effect size .02), that Natural Process (writing and peer collaboration) and Individualized Modes had medium ranges of effect (effect size .19 and .17), and that Environmental Mode (clear specific teacher objectives, strategies, and

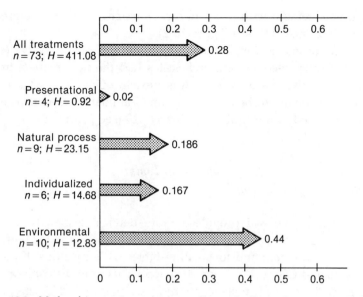

Figure 10-2 Mode of instruction experimental/control effects. *Source:* Hillocks (1983). Courtesy of George Hillocks.

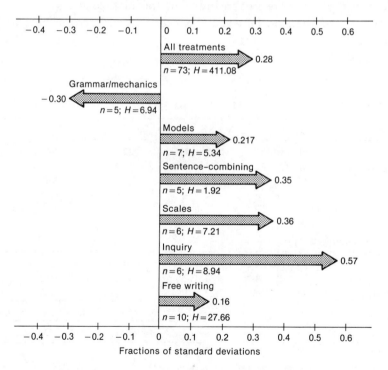

Figure 10-3 Focus of instruction experimental/control effects. *Source:* Hillocks (1983). Courtesy of George Hillocks.

student engagement with problem-centered activities) had the greatest impact (effect size .44) (see Figure 10-2).

For focus of instruction, Inquiry Skills was the most effective (effect size .56); Sentence-combining and Scales had the next highest impact (.35 and .36); the use of Models was next in effectiveness (.22); Free Writing also proved to be effective (.16); Grammar/Mechanics was not only ineffective but harmful (effect size −.29 (see Figure 10-3).

Conducting and reading a meta-analysis: A methodological example

In this section, we will provide a more detailed explanation of the arithmetic methods involved in meta-analysis, as a guide for both conducting such research and for reading tables. (We base this discussion on Hedges and Olkin (1985), pp. 163–165.) This example compares the effect sizes in four sentence-combining studies from Hillocks' meta-analysis with the effect sizes in three sentence-generating studies—one (Faigley, 1979) from Hillocks' meta-analysis and two from his original references which were not included in the meta-analysis. Table 10-3 shows the comparison between mean effect sizes across the two instructional methods.

Let us examine Table 10-3 in detail.

Table 10-3 Data for a Comparative Meta-Analysis

Sentence combining	n_E	n_C	$\tilde{n}$	$\sqrt{MSE}$	d_i	w_i	$d_i w_i$	$d_i^2 w_i$
Howie	51	40	22.42	2.88	.21	22.22	4.66	.98
Morenberg, et al.	151	139	72.38	.95	.34	71.43	24.29	8.26
Pederson	18	18	9.00	.47	.45	8.77	3.95	1.78
Waterfall	19	19	9.50	3.34	.12	9.43	1.13	.13
						$\Sigma w_1 = 111.85$	$\Sigma d_1 w_1 = 34.03$	$\Sigma d_1^2 w_1 = 11.15$
Sentence generating								
Faigley	70	68	34.49	.89	.51	34.48	17.58	8.97
Bond	31	29	14.98	4.84	.87	13.59	11.82	10.29
Hardaway	28	28	14.00	22.22	.13	12.87	1.63	.21
						$\Sigma w_2 = 60.94$	$\Sigma d_2 w_2 = 31.03$	$\Sigma d_2^2 w_2 = 19.47$
						$\Sigma\Sigma w = 172.79$	$\Sigma\Sigma dw = 65.06$	$\Sigma\Sigma d^2 w = 30.62$

Source: Hillocks, 1986, p. 208 (in part).

Note: d_i not corrected from Hedges and Olkin's g since all $n_i > 12$.

READING THE TABLE

$$\boxed{n_e} \text{ and } \boxed{n_c}$$

are the group sizes reported in the original experiment.

$$\boxed{\text{MSE}}$$

the mean square, is also taken from the original experiments. In these experiments the mean square was used for the error term (the denominator of the F test in the Analysis of Variance.) If the results in the experiments are reported as a t test, then the standard deviations (or variances) need to be converted back to their respective sums of squares. This is done by squaring each standard deviation, multiplying each by its respective (n-1), adding these together, dividing this sum by $(n_e + n_c - 2)$, and taking the square root of this quotient, which yields a pooled standard deviation. This pooled standard deviation is then used as the divisor in the formula $(\bar{X}_e - \bar{X}_c)/\sigma_p = $ effect size. Where the mean square for error in an analysis of variance or the standard deviation of the two groups in a t test is not reported, the Appendix of Glass, McGaw, & Smith (1981) gives a series of approaches to help a researcher obtain the necessary standard deviation to compute the d in the following equation in column five of Table 10-3.

$$\boxed{\tilde{n}}$$

is a weighted average if the n's in each group:

$$\tilde{n} = \frac{(n_C)(n_E)}{n_C + n_E}$$

$$\boxed{\sqrt{\text{MS}_E}}$$

is the square root of the mean square described above. Mean squares are really variances and thus their square roots need to be taken to obtain standard deviations, which are used to determine differences between the compared means in the standard score form:

$$d_i = \frac{\bar{X}_E - \bar{X}_C}{\sqrt{\text{MS}_E}}$$

$$\boxed{d}$$

is a z score, a standard score, for showing the differences between the means of the experimental and control groups.

$$\boxed{w_i}$$

allows a researcher to account for varying sizes of the groups used in the different studies. It is used to obtain the weighting for each study in the meta-analysis. Hedges and Olgin (1985 pp. 78–81) give the equation*:

$$w_i = 1\bigg/\left(\frac{1}{\bar{n}_i} + \frac{d_i^2}{2(N-3.94)}\right)$$

$$\boxed{d_i\, w_i}$$

represents d times w.

$$\boxed{d_i^2\, w_i}$$

represents d squared times w.

CALCULATIONS

Table 10-3 shows the results of several calculations. Each of the columns—w_i, $d_i w_i$, and $d_i^2 w_i$—was summed for each of the two groups—sentence-combining and sentence-generating—and then for all seven studies. (For more than two groups, the sums for each of the groups are determined and then added together.) Then a Chi-Square Analysis was done to check for differences between the two groups on Effect Size. The equation used was:

$$\chi^2_1\, df = \frac{(\Sigma\, d_1 w_1)^2}{\Sigma\, w_1} + \frac{(\Sigma\, d_2 w_2)^2}{\Sigma\, w_2} - \frac{(\Sigma\Sigma\, d_T w_T)^2}{\Sigma\Sigma\, w_T}$$

$$= \frac{(34.03)^2}{111.85} + \frac{(31.03)^2}{60.94} - \frac{(65.06)^2}{172.79}$$

$$= 1.70$$

The Chi-Square value was then compared with the tabled value for one degree of freedom (see Appendix I). The results showed no significant differences between these two pedagogies. The effect sizes for each group were then calculated using the following formula:

$$ES_1 = \frac{\Sigma\, d_1 w_1}{\Sigma\, w_1} = \frac{34.03}{111.85} = .30, \ ES_2 = \frac{\Sigma\, d_2 w_2}{\Sigma\, w_2} = \frac{31.03}{60.94} = .51$$

While the averages favored sentence-generating, the difference was not significant. Therefore, the null hypothesis had to be accepted: that

---

*Hedges (1986) drops the 3.94.

sentence-combining has no more effect on writing than sentence-generating.

It was then important to test for homogeneity of Effect Sizes among the seven studies to determine whether any other variables should be considered. The overall H test, using the following formula, was not significant.

$$H_T = \Sigma\Sigma \; d^2 w - \frac{(\Sigma\Sigma \; dw)^2}{\Sigma\Sigma \; d} = 30.62 - \frac{(65.06)^2}{172.79} = 6.10, \text{ N.S. with 6 } df$$

Conducting and reading a correlation: A methodological example

Even though a meta-analysis is able to show comparative strengths among experimental studies, it cannot generate alternative hypotheses as to the reasons why some types of instruction have more effect than others. Researchers, therefore, need to pose alternative hypotheses, searching for significant relationships between other variables and the conclusions in the research studies. Here correlation was used to determine whether a significant relationship existed between instructional time and the Effect Sizes in the seven studies. Below is Table 10-4 and an explanation of its columns.

READING THE TABLE

$\boxed{y}$ represents the instructional time reported in each study.

$\boxed{y^2}$ is y squared.

$\boxed{w}$ is the weight of each study explained above.

Table 10-4 Data for Alternative Hypothesis Testing Via Correlation in Meta-Analysis

	w_i	w_i^2	y_i	y_i^2	$w_i y_i$
Howie	22.22	493.73	1,500	2,250,000	33,330.0
Morenberg, et al.	71.43	5,102.25	1,850	3,422,500	132,145.5
Pederson	8.77	76.91	1,350	1,822,500	11,839.5
Waterfall	9.43	88.92	750	562,500	7,072.5
Faigley	34.48	1,188.87	2,250	5,062,500	77,580.0
Bond	13.59	184.69	1,650	2,722,500	22,423.5
Hardaway	12.87	165.64	540	291,600	6,949.8
	$\Sigma w = 172.79$	$\Sigma w^2 = 7,301.01$	$\Sigma y = 9,890$	$\Sigma y^2 = 16,134,100$	$\Sigma wy = 291,340.8$

Source: Hillocks (1986), p. 208 (in part).

n = 7

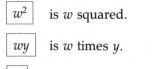

w^2 is w squared.

wy is w times y.

n is the number of studies.

CALCULATIONS

These five terms summed and n were inserted into the correlation formula (See Appendix I, substituting w for X). The correlation coefficient was .58 indicating a non-significant relationship between Instructional Time and the effect size in each study.

Statistical analysis: Descriptive studies

Although most meta-analyses have been done on true and quasi-experiments, which determine cause-and-effect results, they can also be applied to quantitative descriptive research, especially prediction studies. Hunter and his colleagues (1982) have developed a meta-analysis technique for combining or averaging the results of various correlational analyses, taking into account the different conditions of the studies, the degree of random variability of sampling procedures, the unreliabilities of the variables, and the restrictions in the range of subjects. Hunter's techniques can be used if an investigator wishes to generalize the regression weights of various variables such as success in freshman composition courses. Hedges and Olkin (1985, pp. 223–246) have also discussed this use of meta-analysis.

Criticisms and advantages of meta-analysis

Criticisms

Glass, McGaw, and Smith (1981) identify and respond to several criticisms of meta-analysis. The first is that it compares studies done with different measurement techniques, subjects, and so forth. They respond that using only studies that are the same in *all* respects is self-contradictory, because those that need to be compared are those that are different. But if researchers are concerned about this problem, they can decide to use only studies with the same or very similar measures. Hillocks, for example, included only studies that used a scale of quality as a measure of writing, in other words those that used raters of essays. Glass, McGaw, and Smith further argue that this criticism is inconsistent with the common data analysis procedure of lumping "together (average or otherwise aggregate in analyses of variance, t-tests, and whatever) data from different *persons*" (pp. 219–220).

The second criticism is that by including data from "poor" studies, meta-analysis advocates low standards of judging quality (p. 220). Glass, McGaw, and Smith counter by noting that in the psychotherapy studies, on which Smith, Glass, and Miller (1980) did the first meta-analysis, no purpose would have been served by reporting the results of good and bad studies separately because they would have been essentially the same. There was, for example, essentially no correlation between rated quality of research design and effect size of outcome. They suggest that another way of handling this problem, if it is found to affect results, is to use only those studies whose design is strong. Actually, research often has shown that poor studies support the findings of well-done studies, making poor studies more credible (p. 223). However, if poorly designed research results are shown to be different, they can be set aside. Glass, McGaw, and Smith provide a table showing that "as a general rule, there is seldom much more than .1 standard deviation difference between average effects for high-validity and low-validity experiments" (p. 226). A researcher can also use theory to select studies. Hillocks, for example, excluded those studies without a treatment over time and those that did not control for differences of students or teacher bias.

Glass, McGaw, and Smith identify a third criticism—a bias based on whether studies were selected from journals, books, or unpublished papers. They cite evidence that shows that the average experimental effect from studies published in journals is larger than the effect estimated in dissertations—.64 as opposed to .48, respectively. In four of six instances, journals gave more favorable results than books; in four of eight instances, the average effect size in journals was larger than for unpublished studies. While acknowledging the substantive nature of this criticism, they argue that, in fact, meta-analysis is designed to reveal and highlight such biases. They also discuss the bias of time— the period from which the studies are selected. Here, too, evidence reveals that later studies have an average effect size advantage over earlier studies, suggesting that researchers should not arbitrarily select cutoff dates.

Another criticism is that the accuracy of the meta-analysis is influenced by its estimations of complex interdependencies in the data base. To determine the influence of this problem, Glass, McGaw, and Smith offer statistical advice in Chapter 6 of their study of meta-analysis. Slavin (1984, 1986) also offers several critiques of meta-analysis, which are countered by Carlberg et al. (1984).

Advantages

A first advantage of meta-analysis is that it provides a far more rigorous integration of studies than can be done even by the best qualified reviewers of the literature. Second, it can identify gaps in the research literature. Hillocks' research revealed that few studies had been done of the interrelationships among variables within instructional method and curriculum content. Finally, and most important, meta-analysis has value for building theory in a field, because it allows the researcher to generalize broadly over a larger number of variables from many small studies whose conclusions otherwise could only be extended to limited numbers of subjects and situations. Hillocks' work provides the field with more definitive results of instructional modes and focuses of instruction. On the basis of this meta-analysis, composition researchers can conduct more research on the promising modes and focuses, and instructors can do more intelligent curriculum planning. Finally, funding agencies can better allocate funding support for programs. Hillocks noted that the National Institute of Education had recently encouraged traditional mechanics of writing, an irony in the light of his results.

Summary

Meta-analysis is a way of analyzing the combined effect sizes of a number of true and quasi-experiments, and of quantitative descriptive and prediction studies. The studies become the sampling units and the effect sizes become the criterion variables. The researcher then uses correlational analyses, or analysis of variance, for the variables in the study as in any quantitative descriptive study. Meta-analysis has the power to show gaps in a field's literature and to combine numerous small experiments into a major theoretical position. Further, the use of effect size better defines the degree of the importance of variables in a field's theory. Another major advantage is that meta-analysis moves beyond standard tests of statistical significance (with their built-in inability to detect true differences), typical statistical significances in very large samples, and biased "box score" summaries of the literature.

Checklist for reading "Meta-Analyses"

1. What studies were selected as research units? By what criteria were they selected? Were they true or quasi-experiments? Quantitative descriptive studies or prediction studies?
2. What variables were identified as treatment or condition variables for the analysis?
3. What degree of homogeneity did these treatment variables have?
4. What statistical analyses were used?
5. What levels of effect sizes were found?

References

Research Design

Asher, J.W. (1983). Research methodology. In R. J. Corsini (Ed.). *Encyclopedia of psychology*. New York: Wiley.

Bangert-Drowns, R.L. (1986). Review of developments in meta-analytic method. *Psychological Bulletin, 99*, 388–399.

Becker, B.J. (1985). Applying tests of combined significance: Hypotheses and power considerations. Doctoral dissertation, University of Chicago.

Carlberg, C.G. et al. (1984). Meta-Analysis in education: A reply to Slavin. *Educational Researcher, 13*, 16–24.

Cooper, H. (1984). *The integrative research review: A systematic approach*. Beverly Hills Calif.: Sage Publications.

Cooper, H. (1982). Scientific guidelines for conducting integrative research reviews. *Review of Educational Research, 52*, 291–302.

Glass, G.V (1976). Primary, secondary, and meta-analysis of research. *Educational Researcher, 5*, 3–8.

Glass, G.V (1978). Integrating findings: The meta-analysis of research. In L.S. Shulman (Ed.). *Review of research in education*, Vol. 5. Itasca, Ill.: F.E. Peacock.

Glass, G.V, & Smith, M.L. (1979). Meta-analysis of research on the relationship of class-size and achievement. *Evaluation and Policy Analysis. 1*, 2–16.

Glass, G.V, McGaw, B., & Smith, M.L. (1981). *Meta-analysis in social research*. Beverly Hills Calif.: Sage Publications.

Hedges, L.V. (1981). Distribution theory for Glass's estimator of effect size and related estimators. *Journal of Educational Statistics, 6*, 107–128.

Hedges, L.V. (1982a). Estimating effect size for a series of independent experiments. *Psychological Bulletin, 92*, 490–499.

Hedges, L.V. (1982b). Fitting categorical models to effect sizes from a series of experiments. *Journal of Educational Statistics, 7*, 119–137.

Hedges, L.V. (1986). Issues in meta-analysis. In E.Z. Rothkopp (Ed.). *Review of Research in Education*. Washington, D.C.: American Educational Research Association.

Hedges, L.V., & Olkin, I. (1982). Analyses, reanalyses, and meta-analyses. *Contemporary Education Review, 1*, 157–165.

Hedges, L.V., & Olkin, I. (1985). *Statistical methods for meta-analysis*. Orlando, Fla.: Academic Press.

Hedges, L.V., & Olkin, I. (1985). Meta-analysis: A review and a new view. *Educational Researcher, 15*, 14–21.

Hunter, J.E., Schmidt, F.L., & Jackson, G.B. (1982). *Meta-analysis: Cumulative research findings across studies*. Beverly Hills, Calif.: Sage Publications.

Jackson, G.B. (1980). Methods for integrative reviews. *Review of Educational Research, 50*, 438–460.

Light, R.J., & Pillemer, D.B. (1984). *The science of reviewing research*. Cambridge, Mass.: Harvard Univ. Press.

Light, R.J., & Smith, P.V. (1982). Accumulating evidence: Procedures for resolving contradictions among different research studies. *Harvard Educational Review, 41*, 429–471.

Rosenthal, R. (1984). *Meta-analytic procedures for social research*. London: Sage Publications.

Slavin, R.E. (1986). Best-evidence synthesis: An alternative to meta-analytic and traditional reviews. *Educational Researcher, 15*, 5–11.

Slavin, R.E. (1984). Meta-analysis in education: How is it used? *Educational Researcher, 13*, 6–15.

Slavin, R.E. (1984). A rejoinder to Carlberg et al. *Educational Researcher, 13*, 24–27.

Smith, M.L., Glass, G.V, & Miller, T.I. (1980). *The benefits of psychotherapy*. Baltimore, Md.: Johns Hopkins Press.

Meta-Analyses

Hillocks, G. (1984). What works in teaching composition: A meta-analysis of experimental treatment studies. *American Journal of Education, 93*,(1), 133–170.

Hillocks, G. (1986). *Research in written composition*. Urbana, Ill.: National Conference on Research in English.

Stahl, S.A. & Fairbanks, M.M. (1986). The effects of vocabulary instruction: A model-based meta-analysis. *Review of Educational Research, 56*, 72–110.

Waxman, H.C., & Walberg, H.J. (1982). The relation of teaching and learning: A review of reviews of process–product research. *Contemporary Education Review, 1*, 103–120.

Willig, A.C. (1985). A meta-analysis of selected studies on the effectiveness of bilingual education. *Review of Educational Research, 55*, 269–317.

11

Program Evaluations

Program evaluation is not a research design, but rather a type of research that can make use of any of the designs we have discussed. It is often undertaken for administrative or instructional purposes—to determine whether a writing program is achieving its goals, is more effective than an alternative curriculum, is efficiently run, is academically sound, and so forth. There are two basic types of program evaluation: formative and summative. *Formative evaluation* takes place while a program is in operation, sometimes even in stages of development, and is often done at the behest of administrators and those professionals involved with the program. Evaluators examine such features as the curricula, faculty qualifications and attitudes, interim student progress and attitudes, administrative support, environmental factors, teaching methods, kinds of assignments, and types of grading. The object of formative evaluation is to identify strengths and weaknesses while the program is being conducted so that ongoing improvements can be made. Sometimes such evaluation identifies obstacles that are preventing the program from reaching its goals. *Summative evaluation* is concerned with the results of a program, usually in comparison with other programs, and is often done for outsiders. It is conducted not so much to remedy ongoing problems but to make administrative decisions about continuance, faculty effectiveness, and levels of funding.

Discrepancy evaluation (Provus, 1971), a type of both formative and summative evaluation, reviews preset standards of performance on various criteria and actual performances during and at the end of the instructional program. The evaluator uses standards for program implementation and performance that have been set by the director or administrator, matching them against actual personnel, activities, and support. Provus suggests evaluation at five stages: design, installation, process, product, and cost. At any point where a discrepancy occurs, an administrative question is raised and action is considered. When the discrepancy is large, the administrator may have to either institute changes or lower standards. Discrepancy evaluation is thus heavily process-oriented. It also is concerned with educational philosophy, asking what the goals of a program should be.

The need for careful evaluation of writing programs has received much attention in the last several years. With the advent of competency-based outside assessments of students' writing, many writing program directors and English departments have been trying to develop more rigorous ways of conducting their own evaluations. In 1982, a special committee of the Conference on College Composition and Communication (1982) published some guidelines for evaluating college composition programs. It developed and refined these guidelines over several years of discussion with college and university instructors throughout the country. The committee recommended that an evaluation of a composition program should include (1) specification by administrators and instructors of a program's assumptions, goals, and standards for judging writing, (2) self-evaluation of courses by instructors, (3) direct observation of classes, (4) examination of writing assignments, (5) evaluation of teacher's responses to writing, and (6) evalution of instruction by students.

At about the same time, the University of Texas (Witte et al., 1981) received a grant from the Fund for the Improvement of Post-Secondary Education to develop suitable instruments for evaluating writing instruction. To accomplish this difficult task, they engaged in a four-stage process. First, they collected a large pool of statements about composition instruction from writing program directors, students, teaching assistants, and instructors from seven institutions. Second, they constructed a pilot questionnaire containing the items about writing that they had collected. They administered it to students in a large public university, a private four-year institution, and a community college. Third, they gave a revised version of it to 500 students from four institutions, factor-analyzing it and again revising it. Fourth, they tested the revised version with 1552 students, factor-anaylzed it to identify discrete items, and created two versions—a two-part long test (18 and 62 factors) and a 21-item short version. Their report not only provided a helpful example of how to construct an evaluative instrument, but also included the test's scoring uses and a discussion of its limitations.

Davis, Scriven, and Thomas (1981) also provide extensive advice on the evaluation of composition programs. In the second chapter of their book, Scriven explains a number of evaluation concepts, including formative and summative evaluation. He distinguishes between holistic and analytic evaluation, explaining that holistic evaluation reports on how well a program is doing overall, while analytic evaluation offers a diagnosis of the parts that are functioning well or ill. He also discusses six phases of evaluation: (1) previewing the evaluation—planning the design, cost, responsibilities, and so on, (2) developing the evaluation design in detail, (3) conducting it, (4) synthesizing and interpreting results, (5) reporting findings, and (6) evaluating the evaluation (pp. 13–14).

He advises that during an evaluation assessors take several things into consideration. They should consider what elements are to be evaluated (methods, teachers, and student papers?), what kind of evaluation is to be used (formative, summative, holistic, or analytic?), what contextual elements bear on the evaluation (its sponsors, users, intended outcomes), and what resources (time, finances, personnel, etc.) are available (Davis, Scriven, and Thomas, 1981, pp. 14–16). Also important are legal, ethical, and political considerations and constraints. To whom are reports due? To whom can they be released? What records are available for scrutiny (pp. 17–18)? He recommends that evaluators determine the kinds of needs the evaluation is meeting (accountability? proof of effectiveness of a program? improvement of a program?), whether a program is being properly implemented (teacher training, adequate textbooks, pedagogical methods), and what other institutions might benefit from the results of the assessment (pp. 18–20). Also critical are a program's effects—both intended and side effects—on students, their families, and communities, the duration of these effects and their cause-and-effect nature, and a program's generalizability (to whom can its results be extended?) (pp. 21–22). Finally, evaluators must consider the costs of the program and its effectiveness in comparison with alternatives (pp. 22–24).

In Chapter 4 of the same book, Davis and Thomas identify the aspects of composition instruction that should be examined in an evaluation. Because these aspects, as well as the features of an evaluation discussed by Scriven above, are fully delineated in their text, we will only present them briefly here. Evaluators should collect data on all of the following:

1. Student writing performances, which can be assessed by a variety of instruments (pp. 67–95)
2. Student attitudes and beliefs about writing, which can be determined by using observations, self-reports, program records, and surveys (pp. 95–106)
3. Student writing activities and enrollments (pp. 106–108)
4. The process of teaching writing, which can be examined through methodology inventories, questionnaires, structured interviews, classroom observations, students' perceptions of teaching, evaluations of writing assignment topics, and standardized measures of teaching expertise (pp. 109–120)
5. Teacher attitudes and beliefs about writing, gleaned through self-reports, questionnaires, interviews, or reports from others (pp. 120–122)
6. Teacher professional activities and leadership roles (pp. 122–123)
7. Training activities, including in-service training, conference attendance, professional reading determined through logs, inventories, questionnaires, index cards, and evaluations of this training (pp. 123–131)
8. Program administration and costs (pp. 131–142)
9. Unintended side effects and replications (pp. 142–148)

Evaluative designs

Weak designs

Program evaluators have used a variety of designs to conduct evaluations. Witte and Faigley (1983) criticize two of the most common designs, the expert-opinion approach and the quantitative method. They maintain that the limitations of the expert-opinion design are that experts have different criteria, are not capable of evaluating all aspects of a program, often lack knowledge of research principles, and operate largely atheoretically (pp. 6–7). Although Witte and Faigley point out several advantages of the quantitative design over the expert-opinion design such as seeking objectivity in quantitative measures that can be used across evaluation settings (p. 7), publishing results, and using comparisons and measures of changes in students over time—they contend that this design also often operates atheoretically and draws conclusions only valid for the institution for which it is done (pp. 34–35). Other limitations they cite include the fact that quantitative designs rely on products, a practice based on certain inadequate assumptions about writing, its measurement, and the nature of writing classroom activities (pp. 36–37). This design, moreover, in their judgment, does not lead to an assessment of the program's appropriateness for the students served by it, nor does it examine teacher preparedness (pp. 37–38). They advocate an evaluation design that examines five aspects of a program: its cultural and social context, institutional context, structure and administration, content or curriculum, and instruction (pp. 39–65). To investigate these dimensions of a program, they advise using multiple methodologies.

Empirical designs

Many of the research designs that have been explained in the preceding chapters of this book can be used for program evaluation. The choice of design depends on the types of decisions to be made as a result of the evaluation.

DESCRIPTIVE EVALUATION

At the descriptive level, evaluators can conduct their assessment much like a case study or ethnography. They can gather statements about course or program objectives and the materials and instructional methods used to achieve these objectives. They can observe classes, interview or poll instructors and students, and then compare these data in a rather general, informal way with other known similar courses, instructional methods, and materials in order to make decisions about quality and achievements. This kind of evaluation seeks to determine the status of a course or program in the light of the proposed objectives. Evaluators can ask what the desired status should be and what can be done to achieve it. Often descriptive evaluation is *formative* because it provides ongoing information about the implementation of a program. The Davis, Scriven, and Thomas study (1981) illustrates the use of case study design. The expert-opinion evaluation reviewed by Witte and Faigley (1983) is a version of this type of evaluation.

QUANTITATIVE DESCRIPTIVE EVALUATION

As Witte and Faigley (1983) have observed, a more sophisticated model entails administering pretests and posttests to determine whether students have the information or behaviors at the end of the course that had been listed as course objectives. The evaluator can use appropriate standardized achievement tests, observe accomplishments and behaviors, and then decide whether each of the objectives has been met. This model goes beyond the prior one because it asks whether the students knew the material, or much of it, and whether they had the behaviors and skills *before* they started the instructional program. By pretesting and posttesting, observing or rating each criterion variable, the evaluator determines the gains made over the period of instruction. The analysis can involve the simple subtraction of the pretest scores from the posttest scores on all of the criteria variables, which produces a simple gain score on each. (This is an acceptable procedure if no correlational analyses of the gain score variables are to be done.)

Statistical analyses can also determine whether these gain scores are significantly different from a zero gain, but one must remember that when both the pretest and posttest scores are used in the analysis, these scores are *not* independent of one another but are correlated-data or repeated measures. A repeated-measure or correlated-data analysis is more sensitive to possible differences. If evaluators relate the simple subtracted gain scores with other variables such as IQ scores, prior grades in English, or English achievement test scores, they must remember that the correlational results will be biased. When additional analyses are to be made, the true gain scores should be estimated from regression equations (Asher, 1976, pp. 250–252).

This pretest and posttest model, sometimes dubbed "the political model," often gives the course or program credit for all gains that occurred during the time of the instruction. It generally ignores the influences of natural growth, cognitive development, reading, television, home, and other courses. It also ignores gains that are well known to occur from simply retesting (the threat of Testing) and from measurement problems like Regression-Toward-the Mean. This design answers more questions than the descriptive design, but when used alone, it suffers from the limitations discussed by Witte and Faigley.

TRUE EXPERIMENTAL EVALUATION

The strongest evaluation design parallels the true experiment. If evaluators have at least two groups, control and experimental, and can assign students randomly to these groups, then they have a powerful evaluation design on which to base decisions. During or at the end of the course, evaluators collect test results, compositions, questionnaires, and interview data, compare them between the treatment and control groups, and make recommendations for improvement or continution of the program. This evaluation model has some major advantages: no pretests, simpler statistical analysis, and easy interpretation of results. Most important, it enables the evaluator to make cause-and-effect assessments of the relationship between the instructional treatment and the criterion variables.

THE QUASI-EXPERIMENTAL EVALUATION

An evaluation uses a quasi-experimental design if the same or similar pretests and posttests are given to non-randomized program and control groups. If the control groups are similar to the evaluation group initially, then the evaluation is a strong quasi-experimental research design. The researcher can also conduct a general form of weak quasi-experiment, comparing standardized achievement tests scores at the end of a freshman program with the *norms* reported in technical manuals for standardized test scores from similar ages. For example, the evaluator compares the gains between students' posttests at the end of their freshman year and those at the end of their sophomore year with the number of point gains the average normative group in the United States achieves between the end of the freshman year and the beginning of the sophomore year. Because this comparison often is essentially between nonequivalent groups, the evaluation design is like that of a weak quasi-experiment. But it works rather well if the goals of the course or program are similar to the general instructional goals in classes in the United States. If, however, the classes being evaluated are above or below the U.S. average in intellectual ability, the expected ability would fall above or below the norm, requiring a statistical adjustment.

Much of the information about research design and measurement already treated in this text will help in designing and interpreting program evaluations. Knowing the difference between descriptive and experimental research helps to determine when cause-and-effect statements about programs are appropriate. Understanding reliability and validity are important because evaluation often involves systematic measurements. Awareness of the threats to internal validity aids in interpreting apparent changes in variables, in determining the relative importance of relationships among variables, and in using comparison data to explain fully program data on criterion variables. Sampling methods are useful in reducing the cost of data collection or the time involved in observing or interviewing. Comprehending the appropriate use and interpretation of statistics is also essential.

Criteria for evaluation designs

Stufflebeam et al. (1971) suggest a number of criteria for evaluation: relevance, scope, importance, credibility, timeliness, pervasiveness, and efficiency. An evaluation is *relevant* if it gives decision makers adequate responses to questions they want answered. It has the proper *scope* if the breadth of data to be collected is sufficient to describe the program and related variables. It has *importance* if it focuses on the potential value of the data to those who will use it. Its *credibility* rests on the users' belief in the competence of the evaluators—their integrity, diligence, and thoroughness (self-evaluation or evaluators with an interest in the outcome jeopardize credibility). Its *timeliness* is good if the results of the evaluation are available before program decisions have to be made. It is properly *pervasive* when its results are disseminated to all the groups and individuals who need them; further, the failure to disseminate negative results becomes a factor in judging the merit of evaluations. Finally, an evaluation is *efficient* if it takes into consideration the expenses and personnel costs, considering the scope and breadth of the evaluation, a criterion that must be weighed against the importance of the program to be evaluated.

Writing program evaluations

Many writing program evaluations, especially those using the expert-opinion design, are only available to institutions that commission them. For accounts of other designs, Davis, Scriven, and Thomas (1981) can be consulted. They provide several evaluation vignettes and a lengthy account of a case study evaluating State College's "Writing across the Curriculum" program (pp. 149–190). Witte and Faigley provide descriptions and critiques of program evaluations at the University of Northern Iowa, the University of California at San Diego, Miami University, and the University of Texas. White (1985) discusses principles of program evaluation and identifies several specific evaluations, including his own evaluation with Polin (1983) of writing programs on nineteen campuses of the California State University.

Summary

Program evaluation in composition is the process of determining the worth or status of a writing program for several purposes: to provide a basis for making decisions about educational progress, to relate accomplishments to objectives, and to establish the effectiveness of instructional programs. Several evaluation designs can be used— qualitative descriptive, quantitative descriptive, and true and quasi-experimental. Whether doing formative or summative evaluation, the researcher applies measurement procedures that must be reliable, valid, and credible, and from time to time establish distinct cause-and-effect relationships between instructional treatments and outcomes. In assessing the merits of these evaluations, readers can apply practical and design criteria applicable to evaluation. Program evaluation is, in short, a process of becoming thoroughly informed about an instructional program. A reader can ask the following questions of an evaluation.

Checklist for reading "Program Evalutions"

1. What was the purpose of the evaluation—formative or summative?
2. What type of evaluation design was used? Was it appropriate?
3. Were the measurements reliable, valid, and credible?
4. What aspects of composition instruction were included? How were they assessed?
5. What criteria were set? By whom?
6. To whom were the results given?
7. What effects did the evaluation produce?
8. Was the evaluation relevant, important, timely, pervasive, and efficient? Did it have the proper scope?

References

Research Design

Asher, J.W. (1976). *Educational research and evaluation methods.* Boston: Little, Brown.

Bloom, B.S., Hastings, J.T., & Madaus, G.F. (1971). *Handbook on formative and summative evaluation of student learning.* New York: McGraw-Hill.

Borich, G., & Madden, S. (1977). *Evaluating classroom instruction: A sourcebook of instruments.* Reading, Mass.: Addison-Wesley.

Boruch, R.F., & Cordray, D.S. (1980). *An appraisal of educational program evaluation: Federal, state and local agencies.* Evanston, Ill.: Northwestern Univ. Press.

Bursteen, L., & Freeman, H.E. (1985). *Collecting evaluation data.* London: Sage Publications.

CCCC Committee on Teaching and Its Evaluation in Composition (1982). Evaluating instruction in writing: Approaches and instruments. *CCC, 33,* 213–228.

Centra, J. (1977). The how and why of evaluating teaching. In J.A. Centra (Ed.). *Renewing and evaluating teaching.* San Francisco: Jossey-Bass.

Cronbach, L.J., Ambron, S.R., Dornbusch, S.M., Hess, R D., Hrnik, R.C., Phillips, D.C., Walker, D.F., & Weiner, S.S. (1980). *Toward reform of program evaluations: Aims, methods, and institutional arrangements.* San Francisco: Jossey-Bass.

Cook, T., & Reichardt, C. (1979). *Qualitative and quantitative methods in evaluation research.* Beverly Hills, Calif.: Sage Publications.

Costir, F., Greenough, W.T., & Merges, R.J. (1971). Student ratings of college teaching: Reliability, validity, and usefulness. *Review of Educational Research, 20,* 511–535.

Davis, B., Scriven, M., & Thomas, S. (1981). *The evaluation of composition instruction.* Inverness, California: Edgepress.

Fetterman, D.M. (1980) Ethnographic techniques in educational evaluation: An illustration. In A.A. Van Fleet (Ed.). *Anthropology of education: Methods and applications.* [Special Issue] *Journal of Thought, 15*(3), 31–48.

Fitz-Gibbon, C., & Morris, L. (1978). *How to design a program evaluation.* Beverly Hills, Calif.: Sage Publications.

Herman, J. (Ed.) (1987). *Program evaluation kit.* 2nd ed. Beverly Hills, Calif.: Sage Publications.

Lindemann, E. (1979). Evaluating writing programs: What an outside evaluator looks for. *Writing Programs Administration, 3,* 17–24.

Morris, L., & Fitz-Gibbon, C. (1979). *How to measure program achievement.* Beverly Hills, Calif.: Sage Publications.

Patton, M. (1980). *Qualitative evaluation methods.* Beverly Hills, Calif.: Sage Publications.

Provus, M. (1971). *Discrepancy evaluation.* Berkeley, Calif.: McCutchan.

Rossi, P., Freeman, H. & Wright, S. (1979). *Evaluation: A systematic approach.* Beverly Hills, Calif.: Sage Publications.

Rutman, L. (1977). *Evaluation research methods: A basic guide.* Beverly Hills, Calif.: Sage Publications.

Scriven, M. (1967). The methodology of evaluation. In R.W. Tyler (Ed.). *Perspectives on curriculum evaluation.* Chicago: Rand McNally, pp. 39–83.

Smith, D., & Fraser, B. (1980). Toward a confluence of quantitative and qualitative approaches to curriculum evaluation. *Journal of Curriculum Studies, 12,* 367–370.

Stufflebeam, D.L., Foley, W.J., Gephart, W.J., Gubp, E.G., Hammond, H.D., & Provus, M. (1971). *Educational evaluation and decision making.* Itasca, Ill.: E.F. Peacock.

White, E.M. (1985). *Teaching and assessing writing.* San Francisco: Jossey-Bass.

Witte, S., & Faigley, L. (1983). *Evaluating college writing programs.* Carbondale, Ill.: Southern Illinois Univ. Press.

Witte, S., Daly, J., Faigley, L., and Koch, W. (1981). *The empirical development of an instrument for reporting course and teacher effectiveness in college writing classes.* Fund for the Improvement of Postsecondary Education Grant No. G008005896. University of Texas, Austin.

Witte, S., Meyer, P.R., Miller, T.P. and Faigley, L. (1981). *A national survey of college and university writing program directors.* Technical Report No. 2. Fund for the Improvement of Postsecondary Education Grant Nol G008005896. University of Texas, Austin.

Witte, S., Daly, J., Faigley, L., & Koch, W. (1983). An instrument for reporting composition course and teacher effectiveness in college writing programs. *RTE, 17,* 243–262.

Writing Program Administrators Board of Consultant Evaluators (1980). Writing program evaluation: An outline for self-study. *Writing Program Administration, 4,* 23–28.

Program Evaluations

Brown, B., & Harwood, J. (1984). Training and evaluating traditional and non-traditional instructors of composition. *Journal of Basic Writing, 3* (4), 63–73.

Cooper, C., with Cherry, R., Copley, B., Fleisher, S., Pollard, R., & Sartisky, M. (1984).

Studying the writing abilities of a university freshman class: Strategies from a case study. In R. Beach and L. Bridwell (Eds.). *New directions in composition research.* New York: Guilford Press, pp. 19–52.

Lindell, E. (1980). Six reports on free writing: A summary of the FRIS project. *Didakometry.* Malmö, Sweden: School of Education, No. 61.

Teaching and learning the art of composition: The Bay Area Writing Project (1979). *Carnegie Quarterly*, 27(2), 7.

White, E., & Polin, L. (1983). *Research in effective teaching of writing: Final report on phase 1.* NIE-G-81-0011.

Reports of program evaluations

Davis, B., Scriven, M., & Thomas, S. (1981). *The evaluation of composition instruction,* Inverness, California: Edgepress, pp. 203–205.

Witte, S., & Faigley, L. (1983). *Evaluating college writing programs.* Carbondale, Ill.: Southern Illinois Univ. Press, pp. 8–28.

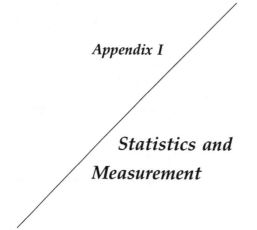

Appendix I

Statistics and Measurement

Our purpose in this appendix is to present the names and purposes of most of the major statistical analyses that are widely used in the behavioral and social sciences. For the reader of empirical research on composition, this background may be useful in interpreting the results of research studies in greater depth. The appendix cannot be and is not intended as a full treatment of statistics. We will refer to texts that we believe to be most accessible to someone without extensive statistical background.

Descriptive statistics

For any variable under scrutiny in empirical research, two concepts are important to understand: central tendencies and individual differences.

Central tendencies

Most data tend to cluster somewhere near the center of scores on a distribution. As scores diverge from the center, they generally become fewer both to the right and left of center. The center of distribution is called the *central tendency*, which has three indicators: the mean, the median, and the mode.

The *mean* $(\bar{X})$ is the average, the sum of the scores divided by their number. The *median* is the middle score, determined by arranging the data scores from lowest to highest in rank order and then counting to the middle. (If there is an even number of scores, halfway between the highest score of the lowest half and the lowest score of the highest half is the estimated median.) The *mode* is the most frequent score, or the data interval that has the most scores. Usually the mode indicates the number of major data points a distribution has: unimodal, bimodal, and so forth.

Data scores that define a variable are generally described as X_i. The number of scores is abbreviated as n or N, n usually being a subset of some N, which is the total number of scores in the set of data. The sum is represented by $+$ or Σ. The mean is the sum of the data numbers divided by their number, $\bar{X}$ or $M = \Sigma X_i/N$.

Individual differences

The degree of individual differences, or the spread of the data, is indicated by the range, the standard deviation, and the variance.

RANGE

The range is the highest score minus the lowest score in a distribution (occasionally defined by purists as the highest score minus the lowest plus one). The range is of limited value because it is based on just two data points and is thus highly variable from sample to sample.

STANDARD DEVIATION

The standard deviation (σ) roughly equals the highest score minus the lowest score divided by five, or range/$5 = \sigma$. For a precise definition, the squares of the scores must be used, X_i^2. Each score is squared. ΣX_i^2 indicates that each individual score is squared and then all of these squared scores are summed, (while $(\Sigma X_i)^2$ indicates that the scores are summed and then this sum is squared).

The formula for standard deviation is a combination of these two terms and n:

$$\sigma = \sqrt{\frac{\Sigma X_i - \dfrac{(\Sigma X_i)^2}{n}}{n-1}}$$

In large samples, scores generally have a range from about −3.00 to +3.00, six standard deviation units in all. In small samples, the range is nearer five standard deviations.

VARIANCE

The variance is another statistic for indicating individual differences. It is simply the standard deviation squared: σ^2. Variance also expresses the extent of individual differences among a set of scores. Many researchers want to know more than whether a relationship between two variables is statistically significant or unlikely to occur by chance. They may also want to know the strength of the relationship in order to determine the amount of variance in a dependent variable that is explained by a given independent variable. The greater the relationship between two variables, the more the independent variable can be said to account for the variance of the dependent variable. If an independent variable does account for a large percentage of the variance of a dependent measure, relative to the total variance, that variable can be said to have a large influence or a strong relationship on the dependent variable.

Two kinds of variance exist: systematic and random. Systematic variance is the variance in a dependent variable that is explained by the influence of an independent variable. Random variance is what is left unexplained, often assumed to have occurred by chance. The researcher wants to reduce that unexplained variance by trying to find other variables that will explain it.

DATA DISTRIBUTION

The *skewness* of a distribution refers to whether there is a "tail" on the distribution to the right or left, called "right skewed" or "left skewed." The *kurtosis* of a distribution describes whether the distribution is unusually fat or thin, flat or peaked, as compared to the normal, bell-shaped distribution.

Standard scores

Because there are a large number of ways in which a psychological construct can be measured, using tests and ratings that differ in terms of number and difficulty of items and judges, a raw score or observational rating usually has little meaning until it is converted into standard scores, the fundamental metric of psychology. A standard score, z, is determined by substracting the mean of the distribution from each score and dividing by the standard deviation of the distribution. These z scores have a range from about -3.00 to $+3.00$.

Most standardized tests meet the goal of producing essentially equal-interval scores because they have been constructed so that when groups of people (like those on whom the test was normed) take the test, their scores can be considered equal-interval data. Often the raw scores on these tests are transformed to standard scores. Because there is no true zero in most psychological data, and any test can consist of a few items or many, an almost infinite number of measurement system scores are possible even on the same dimension of human ability. For example, when measuring general intelligence with a short test, the resulting number of items correct is less than with a medium-length or long test. Thus, it is common practice to convert all psychological measurement systems to a standard-score metric system, z scores, or some scale based on z scores. The standard-score system produces a score of zero for the average raw score on the test. Then it sets one standard deviation of the raw scores as the interval distance. If a raw score is represented by X_i, the transformation to the standard-score system is accomplished by the formula:

$(X_i - \bar{X})/\sigma =$ Standard-score measurement (z scores) where $\bar{X} =$ mean of X.

This formula transforms all raw-score distributions to a common, reasonably comparable standard-score system of z scores, having an average (or mean) of zero and a standard deviation of 1.00. About two-thirds (68%) of most individuals' scores on any given test will fall between the standard scores of minus one standard deviation and plus one standard deviation (± 1 SD). Another 14% will have standard scores between plus one and plus two SD (and also between minus one and minus two). About 2% of test scores will be above the plus two or below the minus two SD points. Almost all of the scores will be between the minus three SD point and the plus three SD point. Figure A-1 shows this distribution.

The important implication of these standard-score units is that they are assumed to be equal-interval measurement units in psychology. They are essentially standard deviation units (functions of the standard deviation), and it has long been known that the standard deviation is proportional to the fundamental measurement interval unit in psychology, the equal-appearing interval in a representative group of people (Jones, 1971).

The minus signs and the decimal parts of the standard deviation units

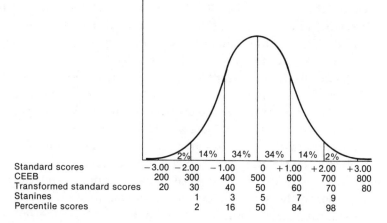

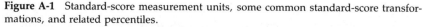

Standard scores	−3.00	−2.00	−1.00	0	+1.00	+2.00	+3.00
CEEB	200	300	400	500	600	700	800
Transformed standard scores	20	30	40	50	60	70	80
Stanines		1	3	5	7	9	
Percentile scores		2	16	50	84	98	

Figure A-1   Standard-score measurement units, some common standard-score transformations, and related percentiles.

make the common standard-score system unwieldy to many people, so this basic standard-score measurement system is itself transformed to four or five other standard-score metrics as indicated in Figure A-1. One of these is the College Entrance Examination Board (CEEB)-type set of scores (e.g., SAT scores), which arbitrarily put the average score at 500 and the standard deviation at 100. Another standard-score metric is one-tenth of the CEEB-type scores with a mean of 50 and a standard score of 10. Both of these metrics eliminate minus scores and decimals. The Stanine score (standard nine) has a nine-point metric with a mean of 5 and a standard deviation of not quite 2.00. All of these variations of the basic standard-score system can be considered good approximations of equal-interval measurement scores.

Percentile scores

The often-used percentile scores are rank-ordered but *not* equal-interval scores, when compared to any of the standard-score metrics. (See Figure A-1 for comparison.) The percentile distance from the mean to the first standard deviation is 34 percentile units, whereas the percentile distance from the first to the second standard deviation unit is only 14 percentile units. The deviation is even greater above the plus two standard score (and below the minus two score), where there are only about two percentile-unit scores from the plus two standard score to the top score. These comparisons of the psychological equal-appearing metric (as reflected in the standard-score system and its transformations) with the percentile rank scores show the distortions of the percentile rank scoring system. This is not much of a problem for scores not too far from the mean of the distribution; however, it does become of concern when working with gifted and talented or remedial populations. The percentile scores that do not seem too far apart at the upper and lower ends of the population are actually rather enormous differences psychologically. Hence, it is better not to use percentile score systems or any rank-ordered data for data analysis.

Effect size and z scores

Effect size is calculated by dividing the differences between the means of the experimental and control group data by the pooled standard deviations. A standard score, z, is derived that shows the effect size, the strength of the treatment group effect as compared with the control group effect. Effect sizes range from about -3.00 to zero to $+3.00$. An effect size of about .20 to .50 is important; anything above .50 is a major difference. Effect sizes for studies can be related to major variables in experimental and control groups.

Inferential statistics

Inferential statistics—unlike descriptive statistics, which are meant simply to describe a given set of data—are used to make inferences from sample groups to what the results of research would be in a population from which the sample was drawn. Researchers postulate the existence of a very large (virtually infinite) set of data, a population. From this large population, they further postulate randomly drawing subsets of data, *on a single variable,* first one sample and then the other, calculating the means of each of the two samples on that variable, and then comparing the size of the difference between these means. If both samples come from the same population, then the average, or *the expected value,* of the difference between the means of the samples thus drawn will be zero.

The compelling interest is more than in the means; it is also in the range or spread of the distribution of the differences between the pairs of means over a large number of pairs of samples. This spread is a function of the range of scores of the original large population and is scarcely affected by the nature of the distribution of the scores in the population, whether it is "normal" or otherwise, as long as the sample sizes are reasonably large, greater than about 30. Researchers ultimately want to learn whether any observed differences that occur in a given pair of samples are rare or common. Researchers determine this using a *t*-test by comparing their actual results with the distribution of results that theoretically would have been obtained simply by randomly drawing many paired sets of samples from a single large population. They are looking for probable or improbable results, not definitive answers. Only the convergence of results over a series of studies moves toward greater certainty about what is happening in the real world. The logic is inferential, from the sample to the population, attempting to characterize a population on the basis of successive samples.

Correlation and regression

Correlation (r), or more specifically the product-moment correlation attributed to Karl Pearson, is an index that measures the strength and type of the relationship between the scores on two variables, X and Y, in a single sample of subjects. The strength of r can vary from no relationship at all (.00) to a perfect relationship (+1.00) to a perfect but inverse relationship (−1.00), where the high scores on one variable are related to the low scores on the other variable.

The basic computational formula for correlation is:

$$r_{xy} = \frac{N(\Sigma XY) - (\Sigma X \, \Sigma Y)}{\sqrt{[N(\Sigma X^2) - (\Sigma X)^2][N(\Sigma Y^2) - (\Sigma Y)^2]}}$$

A *regression equation* is used to predict Y values from X values. In a general form it is

$$\hat{Y} = bX + A$$

where $\hat{Y}$ is the predicted Y score, b is the weight given to the predictor variable X, and A is a constant of adjustment for the two different metrics of the variables X and Y. The b weight is a function of the size of the correlation between the variables X and Y. (If both variables X and Y are scored in standard z-score form, then the b weight is equal to r, the correlation between X and Y, and the constant A is equal to zero.)

If researchers wish to calculate all correlations among all possible pairs of several variables (a matrix of correlations) for a sample of subjects, they correlate the first variable with the second variable, then with the third variable, then with the fourth variable, and so on until the first variable has been correlated with all the variables in the matrix. Then the second variable is correlated with the third variable, and so forth. (Because the second variable has already been correlated with the first, the second variable needs to be correlated only with the third variable, and so on.) Typically, product-moment correlations in the behavioral and social sciences do not reach $+1.00$ or -1.00. In fact, in most circumstances they cannot because of the imprecision of measurement intrinsic to most behavioral and social sciences.

Multiple correlation and regression

It is quite possible to have two or more independent variables used to predict a criterion variable. Usually the predictor variables will be labeled with X's (with various identifying subscripts) and the criterion variable will be Y. (Or the criterion variable is denoted as 1 and the predictors are denoted as 2, 3, 4, 5, 6, etc.) The resulting correlation between all the independent variables combined and the dependent variables is called a multiple correlation. Typically the capital letter R or $R_{1.234}$, . . . is used to differentiate a multiple correlation from the simpler two-variable variety. Different weights for the predictor variables usually produce the maximum multiple correlation and prediction of Y. These weights are called beta weights when all the variables are in standard score form (which is seldom); they are called B weights when the variables are in raw scores, their usual form. These betas and B weights are analogous to the b weight in a simple, two-variable regression equation. When all the variables are in standard scores, all the B weights are equal to the beta weights, the means of all the variables are zero, and no additional adjustment weight (A) is necessary. For almost all practical prediction purposes, though, it is necessary to use a constant A to adjust for the differences in the means of the variables, and B weights are used with each of the raw-score variables. Thus, the assembled multiple-regression equation (with K predictor variables and Y as the criterion variable) will take the form $B_1X_1 + B_2X_2 + B_3X_3 + . . . + B_KX_K + A = \hat{Y}$. While it is possible to do multiple correlation and regression by hand, researchers today use computers for the calculation. The output will typically give both beta and B weights, the A, and the multiple correlation (R) for each set of predictor variables used (or entered).

A regression equation may contain all the available predictor variables, or only certain subsets of variables that predict the criterion variable. These predictor variables can be entered one at a time with the best first, the consequent next best second, and so forth, until R is no longer increased importantly. This is called *forward* entry. It is also possible to enter all the available predictor variables and then take out the least effective predictor first, the consequent second least effective next, and so on until a still important R is achieved with a small number of entered variables. This is called *stepwise backward* entry. It is also possible to get the best predictor set of two, of three, of four variables, and so forth, which can be different from simply the sequentially best two, three, four, etc. variables in a forward or backward procedure. Thus, one must be careful about proclaiming that any given set of predictor variables is the "best" unless all combinations of variables have been investigated.

Beta weights must be used instead of B weights to report the relative importance of each of the predictor variables in a multiple correlation. In a multiple-regression equation the B and A weights are useful only in achieving the best actual predicted score for the criterion variable, given a set of raw-data predictors. (To calculate a B weight for a variable from a beta weight, the standard deviations of the given predictor variable and the criterion variable are used.) A caution must be added. Multiple correlation takes a major advantage of chance relationships when the sample sizes are small and when the ratio of sample size to number of variables is less than 10:1.

Partial correlation

In a partial correlation procedure, the correlation between two variables X and Y is calculated with the effects of a third variable, a fourth, or more held constant or controlled for. It is denoted $r_{12.34}$. . . .

Coefficient of determination and percentage of variance explained

Both r and R can be used to obtain a *coefficient of determination* (r^2 or R^2), which is the percentage of variance in a criterion variable explained by the predictor variables. Thus a correlation between X and Y of $r = .40$, has a coefficient of determination of 16 percent; $r = .40$, $r^2 = .16$, $r^2 \times 100 = 16\%$. As was noted, however, generally 100% of the variance of any variable is not predictable because most variables have reliabilities of less than 1.00. If the reliability of a variable Y is $r = .80$, then only 64% of its variance is explainable. If another variable X has a reliability of $r = .60$, then only 36% of its variance is predictable.

The maximum possible correlation between X and Y, given their reliabilities, is $r_{xy} \leq \sqrt{r_{xx} \times r_{yy}}$ or the maximum for the above example r_{xy} is $\sqrt{.80 \times .60}$, or just less than .70. The maximum percentage of variance explainable between these two variables, X and Y, then is about 49 percent. Thus an actual correlation of $r = .40$ with a coefficient of determination of 16% explains not quite 33% of the explainable variance, $16\%/49\% = 32.6\%$, when the reliabilities of the variables X and Y are taken into account.

Types of correlations

There are several variations of product-moment correlations depending on the types of data being correlated and whether the correlation actually is being calculated or simply estimated.

PHI COEFFICIENT

If both the X and Y variables are dichotomous, the correlation is called a *Phi coefficient* (ϕ). The Phi Coefficient is used to correlate two nominal dichotomous variables like sex (male/female), grade (eighth/twelfth), or age (old/young) that are treated as interval-like variables by labeling one of the two categories as "1" and the other as "2." The results are true product moment correlations derived from the same product moment correlation formula used for two interval variables. Because with Phi the relationship is between two dichotomous variables, Chi Square (see below) can also show this relationship. In fact, the Phi Coefficient and the Chi Square for a four-cell table are closely related,

$$\phi = \sqrt{\frac{\chi^2}{N}}$$

In fact, the Chi-Square statistic is used to test the statistical significance of the Phi Coefficient.

POINT BISERIAL CORRELATION

If only one of the X or Y variables is dichotomous and the other is an ordinary interval variable, the product moment correlation is called a *point biserial correlation*. This type of correlation often is restricted in value to considerably less than $+1.00$, and its tests of significance are different from a traditional correlation. Because this statistic relates a dichotomous nominal and an interval variable, as does the *t*-test (see below), the two statistical test results are identical in terms of statistical significance.

OTHER CORRELATIONS

If both *X* and *Y* have been rank ordered, then the product moment correlation is called a *Spearman Rho*. If both variables either are, or can be considered dichotomized and if the product moment correlation is only being estimated, then the resulting correlation is named a *tetrachoric correlation*. If one variable is dichotomized and the other is interval and the product moment correlation is estimated, the result is a *biseral correlation*. With computers, there is little point to calculating estimated product moment correlations such as the Tetrachoric or the biserial. There is also little point in using special calculational formulas for Phi, Point Biserial, and Rho (to make the computations easier to do by hand). The special correlation names merely reflect the types of variables used for the product moment correlations.

Other types of parametric statistics

An important part of statistics is the *parameter* or the form of the distribution of data for a variable for the population. Parametric statistics are developed under the assumption of a "normal" distribution (i.e., bell shaped). In parametric inferential statistics, the most often used tests are the *t*-test, analysis of variance, and analysis of covariance.

T-*test*

A *t*-test is used when the independent variable is nominal and the dependent variable is interval. A *t*-test determines whether two population means as estimated by two sample means are likely to be the same or different. Often the two samples have been observed after receiving two different treatments or under two different conditions. The *t*-test attempts to determine if there is a true difference between the means of two population groups of sizes N_1 and N_2, based on the data of the sample groups of sizes n_1 and n_2, on a criterion variable X. The formula is simple and uses only terms that have already been developed, except for *degrees of freedom (df)* (see below). The degrees of freedom in this case are equal to $(n_1 - 1) + (n_2 - 1)$.

$$t_{df} = \frac{\bar{X}_1 - \bar{X}_2}{\sqrt{\dfrac{\sigma_1^2}{n_1 - 1} + \dfrac{\sigma_2^2}{n_2 - 1}}}$$

(Note that the sample scores should be independent—that is, not matched, paired, or repeated measures on the same subjects. In the latter case, the correlated *t*-test formula should be used, which adjusts the *t*-statistic for the correlation between the two sets of data.) The *t*-distribution is the theoretical distribution of the difference between two means, $\bar{X}_1$ and $\bar{X}_2$ of the two samples, n_1 and n_2, randomly drawn from the population. The *expected value* (i.e., the average value of the statistic that would be expected from a very large number of random samplings from a single population) of the *t*-statistic given the above circumstances is zero, no difference between the two means. However, it is the variability, the standard deviation, of these possible differences in a large number of pairs of samples that is of primary interest. This standard deviation is simply the denominator of the *t*-test above.

The difference between the two sample means is then compared to the expected variability (the denominator of the *t*-test) to see whether the difference between the means is relatively common or relatively rare. If a relatively common difference is found, then the *t*-statistic is declared not significant; if relatively rare, then the difference is declared significant. In other words, the theorized *null hypothesis* of no difference (when $\bar{X}_1$ is subtracted from $\bar{X}_2$, the result is zero) is either not rejected (not significant) or rejected (significant).

Statistical inference, then, is the process of hypothesizing about what a population is like and the inferences from the statistical data based on samples randomly selected and presumed to be representative of the population. Comparisons between just two means, however, are of primary interest historically as antecedents of the development of analysis of variance.

One-way analysis of variance

One-way *analysis of variance* (ANOVA) is used when one treatment or one condition variable in the research is nominal (but not necessarily dichotomous) and the other criterion variable is interval. The result of an analysis of variance is a statistic, *F*, which is used to test a set of more than two means for possible differences among them. This *F-test* indicates whether or not there is at least one mean among all the nominal categories that is different from one of the others on the interval variable. The statistical null hypothesis is the following: population mean 1 equals population mean 2 equals population mean 3, and so forth. If this null hypothesis is found to be true, the analysis stops. Thus, the analysis of variance tests for possible differences among the means of two or more groups. It also can be used to test just two means; therefore, the *t*-test is technically obsolete, although it is still widely used.

A one-way analysis of variance tests for possible differences on a criterion among several means, as above, or several levels of one treatment. There do not have to be equal numbers of subjects in all the categories of the treatment variable (nor does the *t*-test require equal numbers of subjects).

Two-way and repeated-measures analysis of variance

The real versatility of analysis of variance lies in its ability to test simultaneously two or more treatment variables, each with two or more varying levels in the same analysis. It can also test all the interrelationships, called *interactions,* among the combinations of several treatment variables. Thus the research can more closely approximate the everyday world where several variables regularly operate simultaneously. (It is difficult, however, to analyze and interpret multiple-treatment analysis of variance if the numbers of observations are quite different across the various cells of the combinations of treatment variables.)

Researchers must be careful to make a distinction between analysis of variance using only one observation per subject and repeated-measures analysis of variance (sometimes called correlated-data analysis of variance), which has two or more observations on a single variable per subject (see Edwards, 1985). These analyses are more complex and arise from pre–post tests or repeated observations on the same subject over a period of time on the same variable.

Individual comparisons

Analysis of variance must be used when a researcher wants to determine whether there is a difference among three or more means of a treatment variable. If such a difference is found, all three means could be different from each other, or two of the means on the criterion variable could be equivalent, and the third different from the other two. To specify if more than one mean differs from the others, multiple comparisons are necessary among the means of the levels of a treatment. These further tests in analysis of variance are called *individual comparisons* or *multiple comparisons*. Most of the time, all possible pairs of means of levels of a treatment variable are of interest, but sometimes only comparisons of certain sets of means are of interest. Sometimes these comparisons are all planned in advance of the data collection on the basis of theoretical considerations; at other times the comparisons are made ad hoc. Intermediate and higher level statistics books explain the details of these individual comparison tests, indicating when each should be used (Edwards, 1985; Winer, 1971). Here we will merely list the names of a number of these multiple comparison tests: Fisher's Significant Difference test, the student Neuman–Keuls, the Least Significant Difference Test, Duncan's Multiple-Range Test, Tukey's Significant Difference Test, the Bonferroni Significant Difference Test, the Scheffé Test, and the Protected *t*-Test, Planned Orthogonal Comparisons, Dunnett's Test, and the Honest Significant Difference. Because statisticians continually add to and modify these multiple-comparison tests, the reader of research may encounter additional multiple-comparisons tests in the future.

Error terms in the analysis of variance

In most nonrepeated-measures analyses of variance, the *error term* will be the within-groups mean square in the analysis of variance that is used as the denominator of the *F*-tests to determine the statistical significance of the effects of all the treatment variables and their interrelationships. This is because in most circumstances, all the treatment variables of true and quasi- experiments will be what are described as "fixed effects." With a fixed-effect treatment variable, the levels are not picked at random from among all possible levels of the treatment variable or are proportionately representative of these levels.

The opposite of a fixed-effect variable is a random-effects variable in which the treatment conditions represent a random or proportional representative sample of the population of all possible levels used to represent that treatment. A variable is also deemed a random-effects variable whenever its levels or units are essentially interchangeable or initially indistinguishable from one another, as in the case of subjects. (In the behavioral and social sciences, random-effect variables are deemed unrealistic except as described above.) In these cases, some of the interaction terms of the analysis of variance may be used as the denominator in the *F*-tests to determine their statistical significance. When more than one observation-per-person analysis of variance is used (a repeated-measures analysis of variance), frequently the statistical analysis will also have more than one term that is used as the denominator in the *F*-tests.

The use of fixed-effect and random-effect variables in the analysis of variance is rather complex. Advanced courses in statistics are required to determine what appropriate error terms should be used.

Analysis of covariance

An addition to the analysis of variance, *analysis of covariance* (AN-COVA) uses a covariate that correlates with and adjusts the criterion variable. In true experiments, the main result of the use of a covariant is to reduce the size of the error term, thus making the analysis more powerful. It reduces the probability of a Type II error (see below). In other than randomized experiments, the use of the covariate is used as a control variable to attempt to adjust the criterion variable to take into account or control for possible initial differences among the groups on the covariate. Analysis of covariance can only partially perform this function because of the unreliability of measurement of the variables. Thus caution should be taken in interpreting analyses of covariance for nonrandomized research designs. (The use of multiple covariates is also possible and can be done expeditiously by computers.)

One additional fact and one precaution are needed. The variable used as the covariate must be measured as an interval variable (as are the criterion variables in analysis of variance), because a covariance analysis is essentially a correlational adjustment. The precaution is that the covariate must be observed and measured prior to the application of the treatment. To be useful, the covariate must be correlated with the criterion. If the treatment variable does indeed affect the criterion in a true or quasi-experiment, then because of the correlation, the treatment variable will also affect a covariate measured after the treatment is applied, in much the same manner as the criterion variable. Then, the analysis of covariance, adjusting the criterion with the covariate, tends to remove the influence of the treatment variable on the criterion.

Multivariate data analyses

Multivariate analyses allow the analysis of several dependent variables from one or more samples of individuals. Multiple correlation, already presented, is one type of multivariate analysis. Multivariate analyses are based to some extent on a correlational matrix of a set of K variables. The major types are factor analysis, canonical correlation, multiple correlation, and partial correlation, multivariate analysis of variance, and discriminant analysis. These analyses are almost always used for descriptive research designs in which individuals are observed in their natural environments without the imposition of treatments.

Factor analysis

The main purpose of *factor analysis* (a general term for a family of analytic methods) is to extract an underlying, reduced set of major conceptual variables from a set of real, observed, and interrelated variables on a group of subjects. From a large set of interrelated variables, as in the case of the descriptor variables used by readers in the Diederich study (1974; see Chapter 5), factor analysis sorts variables into clusters that are maximally related to one another, yet relatively independent of all the other clusters of variables. The "cluster" is called a factor. Diederich's factor analysis clustered all the reader's descriptor variables into eight factors.

Factor analysis has many variations including principal components analysis (the most widely used factor analysis procedure), image analysis, alpha factoring, and the older, no longer used Thurston Centroid Method. Factor analysis allows the researcher to talk about major conceptual, underlying dimensions, yet have them each operationally defined by several observational instruments. The factors in a factor analysis can become the general criterion and predictor or treatment variables in a multivariate analysis of variance, in a discriminant analysis, in a canonical correlation, and so forth.

Generally the factors extracted from the correlational matrix are rotated in the multivariate space to a more meaningful pattern that simplifies the interpretation of the factorial structure of the data. The major methods of *rotation of the extracted factors* are the Varimax (the most common), equimax, quartimax, and oblique (in which the factors are allowed to be correlated).

Canonical correlation

Canonical correlation, also used primarily for descriptive research, breaks the correlational matrix among the original variables into two sets: predictor variables and criteria variables. Unlike multiple correlation, which has only one criterion variable, canonical correlation has more than one criterion variable as well as several predictor variables. For instance, one might wish to know the relationship between a set of variables from several multiple-choice tests for punctuation, spelling, usage, and so on, and a set of variables of judges' ratings of style, coherence, and development. A canonical correlation would first factor-analyze the set of interrelations among the predictor set, the multiple-choice test scores, and then factor-analyze the criterion set, the judges' ratings. Then the set of resulting factors in the predictor and criterion groups of measures are maximally correlated, a canonical correlation. New sets of canonical factors are extracted serially from the predictor and criterion sets, but each new canonical factor must have no correlation with all prior canonical factors previously extracted. As researchers learn which variables in the predictor set weigh most heavily with which variables in the criterion set, canonical factor by canonical factor, they will begin to understand how the two separate sets of variables relate to one another and will able to define operationally the crucial variables measured by the tests and ratings. One caution should be given. Canonical correlation requires very large numbers of subjects in order to achieve stable canonical relationships, because for each of the canonical relationships fitted, the degrees of freedom are rapidly reduced. The ten subjects for each variable rule should be a minimum ratio for canonical correlation research.

Multivariate analysis of variance

In one sense, *multivariate analysis of variance* (MANOVA) is any analysis of variance in which more than one treatment variable is used. The two or more treatment variables and their interactions are indeed considered simultaneously with the criterion variable. However, modern usage reserves the term MANOVA to analyses that have multiple criterion variables that are intercorrelated. It is generally not good analytic procedure to perform a separate analysis of variance for each criterion variable without attempting to control Type I error rates. Because the separate results are not independent of each other, their interpretation is clouded. In MANOVA, the criterion set of variables are essentially factored into a lesser number of their fundamental dimensions, which are relatively independent of each other. These resulting criterion factors are then subjected to the usual univariate analysis of variance and interpreted according to which of the original criterion variables contribute the most to the analyzed factors.

Wilks' lambda, one of the multivariate statistics that emerges from the MANOVA, as well as other multivariate statistics, is the multivariate equivalent to the univariate *F*-test for differences among treatment groups. (This is one of the rare statistics that becomes more significant as it becomes smaller.) *Hotelling's T* is the multivariate statistical equivalent of the univariate *t*-test for the MANOVA design with just two treatment groups and several criterion variables.

Discriminant analysis

Discriminant analysis essentially factor-analyzes a set of predictor variables and uses the resulting factors as the predictor variables to predict a group membership criterion variable (a nominal variable). It is similar to a multiple correlation used for a prediction research design with several predictors and a single criterion variable that is measured in intervals. If the criterion variable is nominal—for example, college major, such as English, engineering, or science—then a simple multiple-regression equation should not be used to predict the criterion of college major. The analysis of choice is discriminant analysis, which is sometimes followed by classification procedures using distance functions, such as *Mahalanobis' D.* (For the simple two-criterion group discrimination problems, multiple correlation with a criterion variable scored "1" or "0" is the equivalent of a discriminant analysis and is more easily interpreted.

Nonparametric statistics

Nonparametric statistics are often called distribution-free statistics, and require no assumptions about the nature of the distribution of data in the population. Because nonparametric tests are not quite as powerful in rejecting the null hypothesis as parametric tests, they require somewhat larger samples to be as effective as parametric tests.

Nonparametric statistics often use rank-ordered data and frequency, counted, percentage, and proportion data in nominal categories. (Rank-ordered data analyses, however, generally assume a flat distribution where every rank has an equal frequency.) It is possible to make inferences to a population from a sample of data without regard to the shape of the distribution of the data. These nonparametric statistical analyses are in contrast to parametric statistical analyses, which often assume the normal distribution of a parameter.

Enumeration, counted, or frequency statistics

The *Chi Square* (χ^2) and *"pq"* statistics are the major statistical analyses for enumeration, counted, or frequency data.

1. If the data are simply in raw-score form as the counts of behavior in various subgroups, then Chi Square is used.
2. If the data are in proportions, *"p"* (the counts of data in subgroups divided by the total data count), then *pq* statistics are used.

All summary data are simply *"p"* or *"q"* (which is non-*p*; thus $q = 1 - p$). The mean of enumeration data in proportion form is *p*; its standard deviation is simply $\sqrt{p \cdot q}$. Proportion data may be converted to percentage data by multiplying the proportions by 100, placing both the means and standard deviations in percentage units. Both of these analyses, however, are known as *"pq"* statistics.

Chi Square can be used to show both differences on a variable among groups or relationships among variables within a group, using counted data in the various nominal categories of variables. The null hypothesis is the key in developing the Chi-Square analysis. The researcher essentially asks the question, "What if there is no relationship between these two variables? Or what frequency counts would I expect under such a circumstance? The Chi-Square test then subtracts the expected frequency count per category under this null hypothesis assumption from the frequency count actually observed. These results are squared, summed, and divided by the expected frequency. This Chi-Square statistic is then compared to the distribution of the expected values of Chi Square for the appropriate degrees of freedom. As usual, if the calculated value is equal to or larger than the tabled value of the Chi-Square statistic with the appropriate degrees of freedom, then it is declared significant.

Unlike the parametric statistics, the degrees of freedom used in the simple Chi Square are not based on the number of observations, but usually on the number of categories of a variable less one for one independent variable. For two independent variables such as sex and age, the degrees of freedom are the number of categories (male and female; and ages 0–14, 15–21, 22–30, and over 30) in each variable minus one $(2 - 1$, and $4 - 1)$ and then the multiplication of all these values, $(R - 1)$ $(C - 1)$, $(1 \times 3 = 3)$, and so forth. (Note: R and C represent number of categories in the row and in the column, respectively. See later.)

There are precautions to be observed when using the Chi Square. The observations should be independent (if not, *McNemar's Chi Square* or *Cochran's Q* should be used), and the subjects or their observed behavior must fall in only one category. Care must be taken to assure that the expected number in the several nominal categories is sufficiently large to meet the assumptions of the Chi-Square test. Also note that a Chi-Square test of association of two dichotomous variables is the same as the correlational statistic Phi, which is used as the measure of strength of association for a 2×2 Chi Square (see above). For more than two independent variables or where either of the two variables has more than two categories, the *contingency coefficient (C)* is used as a measure of the strength of the association. *Cramer's Phi*, however, is a better measure of strength of association. Its formula is

$$\phi = \sqrt{\frac{\chi^2}{n(K-1)}}$$

Median and sign tests

The *median test*, a nonparametric analogy of the *t*-test, is actually a Chi-Square test. The *sign test* for matched samples is also a Chi-Square test that is analogous to the parametric *t*-test for matched or paired samples. (Its power to reject the null hypothesis is so low that it should only be used with large samples.)

Log-linear analysis

Log linear analysis transforms enumeration (counted) data into a form more suitable for analysis-of-variance types of statistical analyses, and is highly useful for that purpose. Log-linear analysis is one of the more recently developed methods of statistical analysis.

Rank order statistics

The rest of the nonparametric statistical tests presented here will be rank order tests. The *Spearman* rank order correlation is identified as *rho*. It is a standard product-moment correlation that uses the ranks as equal-interval scores. *Kendall's tau* statistic has the fewest assumptions and is the preferred nonparametric test of a relationship between two variables. Kendall's *coefficient of concordance W*, is the most general coefficient of agreement among a number of judges ordering a set of items or people. The *rank sum* test is the rank order data analogy of the parametric *t*-test. The *Mann–Whitney U* test is another rank order data equivalent of the two-group, parametric *t*-test. (Incidentally, the Mann–Whitney *U* is another of the few statistical tests where small values of *U* are more significant.) With two or more independent samples, the rank order data equivalent of the one-way analysis-of-variance *F*-test is the *Kruskal–Wallis H* test. If the overall *H* test is significant, then a *protected rank sum* test may be used to detect central tendency differences between pairs of groups. The *Wilcoxon test* is the rank order analogy of the matched-pairs two-group *t*-test. The *Friedman* test is an extension of the Wilcoxon for matched individuals for more than two groups.

McNemar's test is a Chi-Square test for two samples, a 2×2 contingency table where the proportions are related or correlated. *Cochran's Q* test is an extension of McNemar's test to more than two related samples.

Concepts used in the interpretation of statistics

Robustness

The ability of any statistical tests on samples of people to yield valid inferences about true differences or relationships in a population is called *robustness*. A statistical test is relatively robust if the test yields reasonably valid results, even when the assumptions such as normality, linearity, or independence used in developing the tests are violated.

Significance and p values

Knowledge in science is probabilistic; knowledge in the behavioral and social sciences is simply more so. Any variable can seemingly differ between groups by chance alone. When this happens, results of statistical analysis are often stated thus: "These two means are not significantly different." What is being stated is that the two means could reasonably differ as much as they do by chance alone. In order to be "significantly different," two means must differ to the extent that it is "improbable" (*not* "impossible") that these differences are due to chance alone.

What are chance differences? Chance differences can be rather precisely defined. They are proportional to individual differences of a variable as expressed by its standard deviation. They are inversely proportional to the square root of the sample size. They are a function of the mathematical expression

$$\frac{\sigma}{\sqrt{n-1}}$$

After chance expectations are defined, social and behavioral scientists next face the problem of developing a decision point on the probability distribution that they will use to distinguish between chance and likely nonchance differences. This point is one at which a researcher will decide that a relationship between two variables exists; and at that point and beyond a chance relationship will be declared improbable (i.e., rare, but again not impossible). If a difference can be called improbable by chance in a sample, then it is more likely to exist in a population. This decision point on the probability distribution is labeled "p." Traditionally, p has been set at .05 and .01, a chance level of one time in twenty (or five times in a hundred) or one time in a hundred. The usual expression is $p < .05$ or $p < .01$ (p "less than" .05 or p "less than" .01), indicating that the observed difference is greater than would be expected by chance at either the $p = .05$ or $p = .01$ level, if the relationship really did not exist.

When statistical tables only recorded values of $p = .05$ and $p = .01$—levels that had been traditional for over half a century—life was simpler. Researchers then used only those levels. (Sir Ronald Fisher strongly influenced the thinking behind recommending that $p = .05$ be the decision point at which the difference between chance and a likely nonchance should result. Sir Ronald, noted for his major ego, is reputed to have once remarked: "How different the world might be if I had set the p value at .20 or .10 rather than at .05.")

The selection of a p value should be based on the status of current knowledge in a field and on the cost of being wrong in one's conclusions. In composition studies, which are relatively young in empirical research, the concern should be to avoid being unable to detect potentially important relationships or differences among variables. Thus some p values for decisions should be set at $p = .10$ and even $p = .20$. Even if one relationship in ten or in five is declared significant when it is not, the error rates are far less than those that occur when relationships are claimed without any empirical testing.

Using p values higher than .05 is no longer difficult, because computer analyses usually report exact p values. If researchers use these p levels, they are less likely to lose important relationships among variables in the background noise created by individual differences, small sample sizes, and less than perfect reliabilities of measurement.

Size of relationship (effect size), rather than statistical significance, is becoming more important in the social and behavioral sciences. Effect sizes, analyzed through meta-analysis, will soon influence theory building more, because nonsignificant results even over a series of studies are known to be rather prevalent, given the usual sample sizes and reliability of measurement in the behavioral science research. Statistical "significance" is only too tenuously related to importance and size of effect of a variable in theory. Finally, with very large samples, statistical significance is not necessarily of practical importance.

One- and two-tailed tests of significance

Once researchers have resolved the problem of an appropriate p level for their research studies, a second problem may arise from time to time in statistical interpretation—whether to use two-tailed or one-tailed tests of statistical significance. Currently in psychological research, the consensus is to use two-tailed statistical tests. Two-tailed statistical significance says that the 5% area of the probability curve is split into two parts with 2.5% being placed at both ends of the distribution. If in a t-test, mean A is *either* markedly greater *or* markedly smaller than mean B, the result will be improbable and fall into one or the other areas of significance and the resulting means will be declared different.

In a one-tailed t-test, the researcher declares that the theory being tested is so clear that only results in which mean A is larger than mean B (or vice versa) are predicted, and the opposite result is meaningless. Thus, these researchers put the whole 5% of the critical area for statistical significance into just one side of the probability distribution. If researchers guess correctly, then the odds are twice as great that they can reject the null hypothesis of no differences. Because, however, most behavioral and social science theory is not that well developed, results in the *opposite* direction are rather embarrassing, especially when critics later point out that they would have been statistically significant. We recommend the widely used current procedure of two-tailed tests of statistical significance. Further, the problem is moot in many statistical tests such as F and Chi Square because the only meaningful tests of significance square differences between the means, thereby losing the sign of the difference. These tables of significance are essentially one-tailed, with the interpretation of direction being done by inspection.

Degrees of freedom

Many statistical tables are entered with the somewhat arcane concept of *degrees of freedom*, which for nominal category data is closely related to the number of categories (male, female; types of majors, etc.), and for interval data is related to the number of observations. The rules for calculating degrees of freedom are rather simple for the more common statistical analyses. (See Tables A-1, A-2, A-3, A-4, and A-5.)

1. For simple Chi-Square analyses of *nominal data,* the degrees of freedom are the number of categories minus one. For a two-variable

Table A-1   Distribution of χ^2

df	Probability	
	.05	.01
1	3.841	6.635
2	5.991	9.210
3	7.815	11.345
4	9.488	13.277
5	11.070	15.086
6	12.592	16.812
7	14.067	18.475
8	15.507	20.090
9	16.919	21.666
10	18.307	23.209
11	19.675	24.725
12	21.026	26.217
13	22.362	27.688
14	23.685	29.141
15	24.996	30.578
16	26.296	32.000
17	27.587	33.409
18	28.869	34.805
19	30.144	36.191
20	31.410	37.566
21	32.671	38.932
22	33.924	40.289
23	35.172	41.638
24	36.415	42.980
25	37.652	44.314
26	38.885	45.642
27	40.113	46.963
28	41.337	48.278
29	42.557	49.588
30	43.773	50.892

Source: Table IV of Fisher & Yates: *Statistical Tables for Biological, Agricultural and Medical Research* published by Longman Group UK Ltd, London (previously published by Oliver and Boyd Ltd, Edinburgh) and by permission of the authors and publishers.

Table A-2   Distribution of t

df	Probability		df	Probability	
	.05	.01		.05	.01
1	12.706	63.657	18	2.101	2.878
2	4.303	9.925	19	2.093	2.861
3	3.182	5.841	20	2.086	2.845
4	2.776	4.604	21	2.080	2.831
5	2.571	4.032	22	2.074	2.819
6	2.447	3.707	23	2.069	2.807
7	2.365	3.499	24	2.064	2.797
8	2.306	3.355	25	2.060	2.787
9	2.262	3.250	26	2.056	2.779
10	2.228	3.169	27	2.052	2.771
11	2.201	3.106	28	2.048	2.763
12	2.179	3.055	29	2.045	2.756
13	2.160	3.012	30	2.042	2.750
14	2.145	2.977	40	2.021	2.704
15	2.131	2.947	60	2.000	2.660
16	2.120	2.921	120	1.980	2.617
17	2.110	2.898	∞	1.960	2.576

Source: Table III of Fisher & Yates: *Statistical Tables for Biological, Agricultural and Medical Research* published by Longman Group UK Ltd, London (previously published by Oliver and Boyd Ltd, Edinburgh) and by permission of the authors and publishers.

Chi Square, the degrees of freedom are the number of categories in the row minus one times the number of categories in the columns minus one, or $(R-1)(C-1)$. (See Table A-1.)

2. For simple analyses with interval data using two groups such as a t-test, the degrees of freedom (df) are the total number of observations in both groups minus two. (For the t-test, two means are calculated, one for each group.) (See Table A-2.)

3. For F-tests two separate degrees of freedom are required to use the F tables to judge statistical significance for the analysis of variance. They are reported in two parts separated by a comma within parentheses, for example, $F_{(1,158)}$. The first degree of freedom, 1, is the value associated with the numerator of the F-test; the second, 158, is the value associated with the denominator of the F-test value. The first value comes from the degrees of freedom for the particular variable's main effect (or interaction), and the second value is associated with the error term in the analysis. The numerator value (1) is found on the top row of the F tables, and the denominator value (158) is found at the left side of the F table in a vertical column. The appropriate F value used for comparison in the table is at the intersection of the appropriate row and col-

Table A-3   5 Percent Points of the F Distribution

df_2 \ df_1	1	2	3	4	5	6	8	12	24	∞
1	161.4	199.5	215.7	224.6	230.2	234.0	238.9	243.9	249.0	254.3
2	18.51	19.00	19.16	19.25	19.30	19.33	19.37	19.41	19.45	19.50
3	10.13	9.55	9.28	9.12	9.01	8.94	8.84	8.74	8.64	8.53
4	7.71	6.94	6.59	6.39	6.26	6.16	6.04	5.91	5.77	5.63
5	6.61	5.79	5.41	5.19	5.05	4.95	4.82	4.68	4.53	4.36
6	5.99	5.14	4.76	4.53	4.39	4.28	4.15	4.00	3.84	3.67
7	5.59	4.74	4.35	4.12	3.97	3.87	3.73	3.57	3.41	3.23
8	5.32	4.46	4.07	3.84	3.69	3.58	3.44	3.28	3.12	2.93
9	5.12	4.26	3.86	3.63	3.48	3.37	3.23	3.07	2.90	2.71
10	4.96	4.10	3.71	3.48	3.33	3.22	3.07	2.91	2.74	2.54
11	4.84	3.98	3.59	3.36	3.20	3.09	2.95	2.79	2.61	2.40
12	4.75	3.88	3.49	3.26	3.11	3.00	2.85	2.69	2.50	2.30
13	4.67	3.80	3.41	3.18	3.02	2.92	2.77	2.60	2.42	2.21
14	4.60	3.74	3.34	3.11	2.96	2.85	2.70	2.53	2.35	2.13
15	4.54	3.68	3.29	3.06	2.90	2.79	2.64	2.48	2.29	2.07
16	4.49	3.63	3.24	3.01	2.85	2.74	2.59	2.42	2.24	2.01
17	4.45	3.59	3.20	2.96	2.81	2.70	2.55	2.38	2.19	1.96
18	4.41	3.55	3.16	2.93	2.77	2.66	2.51	2.34	2.15	1.92
19	4.38	3.52	3.13	2.90	2.74	2.63	2.48	2.31	2.11	1.88
20	4.35	3.49	3.10	2.87	2.71	2.60	2.45	2.28	2.08	1.84
21	4.32	3.47	3.07	2.84	2.68	2.57	2.42	2.25	2.05	1.81
22	4.30	3.44	3.05	2.82	2.66	2.55	2.40	2.23	2.03	1.78
23	4.28	3.42	3.03	2.80	2.64	2.53	2.38	2.20	2.00	1.76
24	4.26	3.40	3.01	2.78	2.62	2.51	2.36	2.18	1.98	1.73
25	4.24	3.38	2.99	2.76	2.60	2.49	2.34	2.16	1.96	1.71
26	4.22	3.37	2.98	2.74	2.59	2.47	2.32	2.15	1.95	1.69
27	4.21	3.35	2.96	2.73	2.57	2.46	2.30	2.13	1.93	1.67
28	4.20	3.34	2.95	2.71	2.56	2.44	2.29	2.12	1.91	1.65
29	4.18	3.33	2.93	2.70	2.54	2.43	2.28	2.10	1.90	1.64
30	4.17	3.32	2.92	2.69	2.53	2.42	2.27	2.09	1.89	1.62
40	4.08	3.23	2.84	2.61	2.45	2.34	.218	2.00	1.79	1.51
60	4.00	3.15	2.76	2.52	2.37	2.25	2.10	1.92	1.70	1.39
120	3.92	3.07	2.68	2.45	2.29	2.17	2.02	1.83	1.61	1.25
∞	3.84	2.99	2.60	2.37	2.21	2.10	1.94	1.75	1.52	1.00

Source: Table V of Fisher & Yates: *Statistical Tables for Biological, Agricultural and Medical Research* published by Longman Group UK Ltd, London (previously published by Oliver and Boyd Ltd, Edinburgh) and by permission of the authors and publishers.

umn. Often several values are found there, one for each p value. The researcher uses the critical F value associated with the previously chosen p value. (See Tables A-3 and A-4.) For a simple *analysis of variance* with four groups, the numerator df is the number of

Table A-4 1 Percent Points of the F Distribution

df_2 \ df_1	1	2	3	4	5	6	8	12	24	∞
1	4052	4999	5403	5625	5764	5859	5982	6106	6234	6366
2	98.50	99.00	99.17	99.25	99.30	99.33	99.37	99.42	99.46	99.50
3	34.12	30.82	29.46	28.71	28.24	27.91	27.49	27.05	26.60	26.12
4	21.20	18.00	16.69	15.98	15.52	15.21	14.80	14.37	13.93	13.46
5	16.26	13.27	12.06	11.39	10.97	10.67	10.29	9.89	9.47	9.02
6	13.74	10.92	9.78	9.15	8.75	8.47	8.10	7.72	7.31	6.88
7	12.25	9.55	8.45	7.85	7.46	7.19	6.84	6.47	6.07	5.65
8	11.26	8.65	7.59	7.01	6.63	6.37	6.03	5.67	5.28	4.86
9	10.56	8.02	6.99	6.42	6.06	5.80	5.47	5.11	4.73	4.31
10	10.04	7.56	6.55	5.99	5.64	5.39	5.06	4.71	4.33	3.91
11	9.65	7.20	6.22	5.67	5.32	5.07	4.74	4.40	4.02	3.60
12	9.33	6.93	5.95	5.41	5.06	4.82	4.50	4.16	3.78	3.36
13	9.07	6.70	5.74	5.20	4.86	4.62	4.30	3.96	3.59	3.16
14	8.86	6.51	5.56	5.03	4.69	4.46	4.14	3.80	3.43	3.00
15	8.68	6.36	5.42	4.89	4.56	4.32	4.00	3.67	3.29	2.87
16	8.53	6.23	5.29	4.77	4.44	4.20	3.89	3.55	3.18	2.75
17	8.40	6.11	5.18	4.67	4.34	4.10	3.79	3.45	3.08	2.65
18	8.28	6.01	5.09	4.58	4.25	4.01	3.71	3.37	3.00	2.57
19	8.18	5.93	5.01	4.50	4.17	3.94	3.63	3.30	2.92	2.49
20	8.10	5.85	4.94	4.43	4.10	3.87	3.56	3.23	2.86	2.42
21	8.02	5.78	4.87	4.37	4.04	3.81	3.51	3.17	2.80	2.36
22	7.94	5.72	4.82	4.31	3.99	3.76	3.45	3.12	2.75	2.31
23	7.88	5.66	4.76	4.26	3.94	3.71	3.41	3.07	2.70	2.26
24	7.82	5.61	4.72	4.22	3.90	3.67	3.36	3.03	2.66	2.21
25	7.77	5.57	4.68	4.18	3.86	3.63	3.32	2.99	2.62	2.17
26	7.72	5.53	4.64	4.14	3.82	3.59	3.29	2.96	2.58	2.13
27	7.68	5.49	4.60	4.11	3.78	3.56	3.26	2.93	2.55	2.10
28	7.64	5.45	4.57	4.07	3.75	3.53	3.23	2.90	2.52	2.06
29	7.60	5.42	4.54	4.04	3.73	3.50	3.20	2.87	2.49	2.03
30	7.56	5.39	4.51	4.02	3.70	3.47	3.17	2.84	2.47	2.01
40	7.31	5.18	4.31	3.83	3.51	3.29	2.99	2.66	2.29	1.80
60	7.08	4.98	4.13	3.65	3.34	3.12	2.82	2.50	2.12	1.60
120	6.85	4.79	3.95	3.48	3.17	2.96	2.66	2.34	1.95	1.38
∞	6.64	4.60	3.78	3.32	3.02	2.80	2.51	2.18	1.79	1.00

Source: Table V of Fisher & Yates: *Statistical Tables for Biological, Agricultural and Medical Research* published by Longman Group UK Ltd, London (previously published by Oliver and Boyd Ltd, Edinburgh) and by permission of the authors and publishers.

groups minus one; the denominator df is the total number of subjects in all groups minus one for each group mean, or N minus the number of groups. Thus the degrees of freedom for the resulting F-test would be $F_{(3, N-4)}$.

Table A-5 Values of the Correlation Coefficient for Different Levels of Significance

	Probability			Probability	
df	.05	.01	df	.05	.01
1	.99692	.999877	16	.4683	.5897
2	.95000	.990000	17	.4555	.5721
3	.8783	.95873	18	.4438	.5614
4	.8114	.91720	19	.4329	.5487
5	.7545	.8745	20	.4227	.5368
6	.7067	.8343	25	.3809	.4869
7	.6664	.7977	30	.3494	.4487
8	.6319	.7646	35	.3246	.4182
9	.6021	.7348	40	.3044	.3932
10	.5760	.7079	45	.2875	.3721
11	.5529	.6835	50	.2732	.3541
12	.5324	.6614	60	.2500	.3248
13	.5139	.6411	70	.2319	.3017
14	.4973	.6226	80	.2172	.2830
15	.4821	.6055	90	.2050	.2673
			100	.1946	.2540

Source: Table VI of Fisher & Yates: *Statistical Tables for Biological, Agricultural and Medical Research* published by Longman Group UK Ltd, London (previously published by Oliver and Boyd Ltd, Edinburgh) and by permission of the authors and publishers.

4. For *correlation* (r_{12}), the degrees of freedom are based on the number of pairs of observations minus one for the mean of each of the two variables, $n - 2$. In simple correlation, one mean is calculated for each of the two variables involved. (See Table A-5.)

Typically researchers compare the results of their statistical tests (with the appropriate degrees of freedom) with the value found in the statistical tables for the level of p that they have selected. The general rule (with only rare exceptions) is that one's obtained results of the statistical analyses must be *larger* than the tabled values for statistical significance to be declared; in other words, the results are nonchance.

Modern researchers with statistical analyses problems of any size will use computers and standard statistical analysis programs such as SPSS, SPSS X, SAS, or BMDP to do their analyses. Computers do not deign to use the tabled 5% and 1% values for the various statistics as mere humans must. Instead, computers insert the calculated data summary statistical values into the rather complex equations defining the normal, t, F, and Chi-Square distributions and calculate the precise value of p for the entered values for that set of data. Thus, when one reads in a journal article $p = .45$ or $p = .06$, one can infer that the analysis was done by computer.

Type I errors

Type I errors occur when a researcher concludes that an observed relationship in a sample is significant which in reality in the population is not—but is instead simply a result of chance. In other words, the Type I error happens when statistically significant differences among the samples occur by chance alone (meaning improbable by chance results). In reality, the difference does not exist, but the researcher interprets it as "significant." These Type I errors are tightly controlled in general practice, and will occur only one time in a hundred or one time in twenty, depending on whether the $p = .01$ or the $p = .05$ level of significance was used. After several studies are done, these errors can be rectified because the extent of the relationship will become clearer. Chance differences will balance out and true differences will become more apparent. Over even three studies, the chance of three results on the same variable being significant by chance three times at the $p = .01$ level is less than one in a million. If researchers want to avoid this error of calling a relationship significant when it really is not, then they should set the p level low, for example, at $p = .01$ or even $p = .001$.

Type II errors

Type II errors occur when researchers are unable to detect true differences or relationships. These errors of statistical inference about data occur when the "noisy" human variability, the imprecise measurement systems, and small numbers of subjects, combine to produce a large natural variability that interferes with the detection of differences and relationships that are truly there. To avoid this error of declaring no statistically significant difference when, in fact, there is a difference or a relationship, researchers can set the p value at a larger value than the traditional ones, for example at $p = .10$ or $p = .20$. These would be respectable values, especially in exploratory research when an investigator has a small sample and relatively unreliable measures. Type II errors cause researchers to overlook valuable relationships among variables because the p value was set too low. These errors are common in behavioral and social science research.

Problems of measurement in research

Many unusual phenomena, like Regression-Toward-The-Mean, seemingly occur in the behavioral and cognitive sciences because their variables cannot be measured with the precision possible in the realms of the physical and biological sciences. (This is one of the principles stated in Chapter 1.) Below are some problem areas in measuring variables.

Reliability

Reliability is agreement among independent observers, or agreement between two alternative forms of a test, or agreement among items in a test. The concept of reliability can be expressed quantitatively as either a percentage of agreement among observers when the variable is nominal as in the percentage of counts that agree in several categories, or as correlation, when the variables are interval scores.

PERCENTAGE OF AGREEMENT

Percentage of agreement is a fairly straightforward process. Two or more coders independently read and assign data or units to two or more categories. Then the number of codings on which they agree are counted and divided by the total number of units. A percentage of agreement entails concurrence by the coders about the set of categories and the assignment of units to them. This percentage of agreement provides some degree of "intersubjectivity," which indicates the reliability of these observers' judgments.

COHEN'S KAPPA

There is a slight flaw in the simple approach to percentage of agreement among observers: a percentage of that agreement may be due to chance alone. Given two categories equally used by both observers, agreement by chance alone will be fifty percent. To address this problem, Cohen (1960) developed a statistic, kappa, to indicate the percentage of agreement that remains after agreement by chance has been removed. In terms of frequencies (counts or numbers), Cohen gives the following formula:

$$\text{Kappa} = \frac{f_o - f_c}{N - f_c}$$

where f_o is the observed frequency of agreement, f_c is the frequency of chance, and N is the number of units that have been coded (Cohen, 1960, p. 40). Kappa has a number of valuable properties. When the obtained percentage agreement equals the chance agreement, kappa equals zero. Obtained percentage agreement greater than chance agreement produces positive values of kappa and with perfect percentage agreement among coders, kappa equals $+1.00$.

The equation for kappa itself is relatively simple. The possibly difficult part is determining what the frequencies of agreement by chance would be. In general, this frequency by chance is the number of units to be coded times the sum of the percentages of each coder's judgments in each of the possible categories. This sum of percentages is then multiplied by the percentages within categories for each of the coders. For instance, if there are three possible categories of judgment and two judges, then each judge will have a number of judgments in each category. Here is an example:

		Coder A			
		Category I	Category II	Category III	Sum
	Category I	30	5	5	40
Coder B	Category II	15	10	5	30
	Category III	15	5	10	30
	Sum	60	20	20	100

If these coding category numbers are 60, 20, and 20 for the first coder (100 judgments in all) and 40, 30, and 30 for the second coder, then the frequencies by chance will be 100 times:

$$\left[\left(\frac{60}{100}\right)\left(\frac{40}{100}\right) + \left(\frac{20}{100}\right)\left(\frac{30}{100}\right) + \left(\frac{20}{100}\right)\left(\frac{30}{100}\right)\right]$$

or $100[.24 + .06 + .06] = 100[.36] = 36$

(which also equals the percentage agreement by chance alone, because $N = 100$.) Kappa can be readily obtained. Observed frequency of agreement equals $30 + 10 + 10 = 50$. Chance frequency of agreement, as calculated, equals 36.

$$\text{Kappa} = \frac{50 - 36}{100 - 36} \text{ or } \frac{14}{64} = .22$$

Ciminero, Calhoun, and Adams (1977, p. 312) suggest that when distributions of obtained frequencies meet the requirements of parametric statistics, Pearson product-moment correlations between pairs of observers can be used as the measure of reliability.

Extensions of kappa for more than two coders, estimates of kappa's variance, statistical tests, and so forth, are available in Fleiss (1971).

CRONBACH'S ALPHA

Cronbach's alpha is the general index of reliability. Procedures for calculating alpha are widespread and found in many standard statistical packages for computers. The older Kuder–Richardson 20 (KR 20) was used for nominal data, which are exclusively scored as right or wrong. This is now no longer needed because Cronbach's alpha encompasses both nominal and interval data.

Reliability coefficients also indicate the precision of measurement instruments. Reliability depends on the number of items in a test or the number of persons coding, grading, or rating a composition. Because no test can have an infinite number of items nor can ratings be made by an indefinite number of observers, all observed correlations understate the true relationship between any two variables in the conceptual world (unless the true relationship is zero). It is thus safe to say that most all observed correlations among composition variables represent something less than the relationships among them in the conceptual world. Further, because of lack of perfect precision of measurement, all observed scores are biased (except those at the mean) and all standard deviations and variances are larger than the true individual differences.

SPEARMAN–BROWN FORMULA

Just after the turn of the century, Spearman and Brown realized that the precision of the social and behavior sciences was related to the number of items in a test or the number of observers making a judgment. They developed an equation to depict that fact:

$$r_{KK} = \frac{K r_{ii}}{1 + (K-1) r_{ii}}$$

where:

r_{ii} is the observed internal-consistency reliability coefficient (ii suggests the relationship of a variable with itself);

K is the factor (or proportion) by which the number of items in a test or the number of observers is increased (or decreased) from the number of items or observers on which the reliability was calculated;

r_{KK} is the projected reliability of the measurement procedure when items or observers are increased (or decreased) by the factor of K.

Because no psychological measurement can ever have a perfect reliability (an instrument of infinite length), the consequences of this imprecision need to be considered.

THE CORRECTION FOR ATTENUATION

The internal-consistency reliability of variables limits the possible degree of correlation between them. The equation used to correct for the unreliability of the variables is

$$\hat{r}_{12} = \frac{r_{12}}{\sqrt{r_{11} \times r_{22}}}$$

where:

- r_{12} is the observed correlation between variables one and two;
- r_{11} is the reliability of the first variable;
- r_{22} is the reliability of the second variable;
- $\hat{r}_{12}$ is "corrected" or an estimate of the true correlation between the variables.

The maximum correlation, then, between two variables is related to their reliabilities. The relationship between the true and observed correlation is called the *correction for attentuation*, which has been used periodically over the years to depict further the relationship between variables. (This correction for attentuation has also been often criticized because it has been incorrectly applied by novices.)

PROBLEMS WITH CRITERION AND PREDICTOR VARIABLES

Adding several criterion variables together to form a single criterion can be a problem. When variables are simply combined, they are self-weighting; in other words, they add as functions of their standard deviation and their intercorrelations. Technically, the variance of a sum of variables is the sum of their individual variances, plus two times the sum of the correlations among the variables, each multiplied by their standard deviations. [See Nunnally (1978) for the rules for combining psychometric variables.]

Care should be taken when adding, multiplying, subtracting, or dividing variables. Researchers can create statistical artifacts that are interpreted as real. Dividing or subtracting variables without relating them by means of regression analyses can be a major problem. When researchers correlate variables that themselves are the result of subtractions, ratios, and quotients, strange results occur. Signs of correlations of two variables will often be negative with a third variable when the subtrahend or divisor variable is positively related to the third variable.

PROBLEMS WITH CORRELATIONS

1. Curvilinear Relationships

A standard assumption in correlational analysis is a straight-line (linear) relationship between the two variables being correlated. Usually this assumption of linearity is approximately correct, but it does pay to look at the plots of the relationship, the scatter diagram. This can now be done readily as part of the standard computer analysis output. If the relationship is curvilinear, the correlation coefficient will understate the actual degree of the relationship. (This is a minor problem, however, at this stage of research in composition.) There is a correlation, labeled *eta*, which maximizes the relationship between two variables, x and y, regardless of the nature of the relationship between them: quadratic, cubic, or general curvilinear.

2. The Split in Dichotomous Variables

Dichotomous nominal variables (e.g., gender) restrict the values of variables to 0 and 1 (or the equivalent). A split that is other than 50 percent restricts the size of the correlational relationship of a dichotomous variable with other variables. A better procedure is to compare the means of the groups represented by the nominal variable with a *t*-test or ANOVA.

3. The Correlations of Averages of Variables over Groups

When a researcher aggregates data in nominal classifications and correlates the averages of these data, the correlations tend to be high, distorting the estimate of what the same correlations would be for the individual's scores on these same variables within the group. Essentially, these averages (or means) based on the sum of scores of a number of people, are more stable than an individual's score. The increased reliability of the summation procedure (via the Spearman–Brown) tends to produce these higher correlations. For example, the correlation of two test scores averaged across classes might be .85 while the individuals' two test scores correlated within the classes might be on the order of .50.

4. The Use of Percentile Scores or the Use of Rank Orders as Equal Intervals

The use of percentile scores or the quantification of variables by means of rank orders distorts the fundamental equal-interval metric of psychology and thus tends to distort the correlations among the true psychological variables.

5. Restrictions in the Range of a Variable

Restrictions in the range of a variable (such as using as subjects only students admitted to a selective college) reduces the observed correlations from those calculated on the total, unrestricted sample.

Assumptions in statistics: Some considerations

Central limit theorem and Tchebycheff's inequality

These two concepts hold for all distributions for which means and standard deviations can be calculated. The *Central Limit Theorem* essentially states that the distribution of a number of sample means approaches a normal distribution and has a variance of σ^2/n as n increases. When the n of the samples is large (more than thirty), the sampling distribution of these means is approximately normal (bell shaped) no matter what the distribution of the original data. Most statistical procedures assume distributions to be normal.

Tchebycheff's inequality

Tchebycheff's Inequality states that the probability of a standardized z score, drawn at random from any data set regardless of the distribution form, occurs with a p value less than or equal to (in either direction) $1/K^2$, where K is the standard score. Clearly, if one has a standard score of 2.50, the probability of finding a score of greater or less than 2.50 standard deviations from the mean is no more than $(1/2.5)^2$, or $1/6.25$, or $p = .16$. Further, if one simply assumes that the distribution has but a single mode and is symmetric, this probability is reduced by a factor of $4/9$. Thus, for the above standard score, the probability with these assumptions is $(4/9) \cdot (1/6.25)$, or $4/56.25$, or $p = .071$. The major value of Tchebycheff's Inequality and the Central Limit Theorem is to show that the probabilities of the statistics calculated assuming a normal distribution for the inferential statistics will not be too far from the actual probability no matter what the distribution of the data in hand.

References

Asher, J.W. (1976). *Educational research and evaluation methods.* Boston: Little, Brown.

Ciminero, A.R., Calhoun, K.S., and Adams, H.E. (Eds.). (1977). *Handbook of behavioral assessment.* New York: John Wiley & Sons.

Cohen, J.A. (1960). A coefficient of agreement for nominal scales. *Education and Psychological Measurement, 20,* 37–46.

Cooley, W.W., & Lohnes, P.R. (1971). *Multivariate data analysis.* New York: Wiley.

Diederich, P. (1974). *Measuring growth in English.* Champaign, Ill.: National Council of Teachers of English.

Edwards, A.L. (1985). *Experimental design in psychological research,* 5th edition. New York: Harper & Row.

Fisher, R.A., & Yates, I. (1953). *Statistical tables for biological, agricultural, and medical research.* 4th ed. London: Oliver and Boyd.

Fleiss, J.L. (1971). Measuring nominal scale agreement among many raters. *Psychological Bulletin, 76*(5), 378–382.

Folger, J.F., Hewes, P., & Poole, M.S. (1981). Coding social interaction. In B. Dervin & M.J. Voight (Eds.). *Progress in communication sciences.* Norwood, N.J.: Ablex.

Guilford, J.P., & Fruchter, B. (1973). *Fundamental statistics in psychology and education,* 5th edition. New York: McGraw-Hill.

Hayes, W.L. (1981). *Statistics,* 3rd edition. New York: Holt, Rinehart & Winston.

Hedges, L.V., & Olkin, I. (1985). *Statistical methods for meta-analysis.* Orlando, Fla.: Academic Press.

Hopkins, K.D., & Glass, G.V (1978). *Basic statistics for the behavioral sciences.* Englewood Cliffs, N.J.: Prentice-Hall.

Hunter, J.E., Schmidt, F.L., & Jackson, G.B. (1982). *Meta-analysis: Cumulating research findings across studies.* Beverly Hills, Calif.: Sage Press.

Jones, L.V. (1971). The nature of measurement. In R.L. Thorndike (Ed.). *Educational measurement.* Washington, D.C.: American Council on Education.

Lindquist, E.F. (1953). *Design and analysis of experiments in psychology and education.* Boston: Houghton Mifflin.

Nie, N.H., Hull, C.H., Jenkins, J.G., Steinbrenner, K., & Bent, D.H. (1975). *Statistical package for the social sciences* (SPSS), 2nd edition. New York: McGraw-Hill.

Norusis, M.J. (1986). *SPSS Guide to Data Analysis.* Chicago, Il.: SPSSFNC.

Nunnally, J.C. (1978). *Introduction to psychological measurement.* New York: McGraw-Hill.

Winer, B.J. (1971). *Statistical principles in experimental design,* 2nd edition. New York: McGraw-Hill.

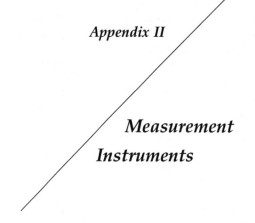

Appendix II

Measurement
Instruments

General lists

Bridgeford, N.J. (1981). *A directory of writing assessment consultants.* Portland, Or.: Clearinghouse for Applied Performance Testing, Northwest Regional Educational Laboratory.

Buros, O.K. (ed.). (1985). *English Tests and Reviews.* Mental Measurements Yearbook Monographs. Highland Park, N.J.: Gryphon Press.

Fagan, W.T., Cooper, C.R., and Jensen, J.M. (1975). *Measures for research and evaluation in the English language arts.* Vol. 1. Urbana, Ill.: National Council of Teachers of English.

Fagan, W.T., Jensen, J.M., & Cooper, C.R. (1985). *Measures for research and evaluation in the English language arts.* Vol. 2. Urbana, Ill.: National Council of Teachers of English.

Mitchell, James (1983). *Tests in print III.* Buros Mental Measurements Yearbooks. Highland Park, N.J.: Gryphon Press.

Stiggins, R.J. (1981). *A guide to published tests of writing proficiency.* Portland, Oregon: Clearinghouse for Applied Performance Testing, Northwest Region Educational Laboratory.

Sweetland, R.C., & Keyser, D.J. (1983). *Tests.* Kansas City, Mo.: Test Corporation of America.

SPECIFIC TESTS

The measurement instruments listed below are clustered according to the aspect of writing or writing behavior evaluated. Where possible, the author and publisher of the measurement instrument are provided. In other cases reference is given to a study that employed the measurement instrument.

1. Attitude and Apprehension Measures

ATTITUDE INVENTORY. Refer to Lague, C.F., & Sherwood, P. (1978). Teaching monograph—co-designed laboratory approach to writing. Doctoral dissertation, University of Cincinnati.

CATEGORIES OF PERSONAL INFORMATION. Refer to Brand, A.G. (1979). The therapeutic benefits of free or informal writing among selected eighth-grade students (Doctoral dissertation, Rutgers University). *DAI, 40,* 1316A–1317A.

COOPERSMITH SELF-ESTEEM INVENTORY. (S. Coopersmith, Palo Alto, Calif.: Consulting Psychologists Press.) Refer to Sussman, G. (1973). The effects of writing about self-on the self-esteem of fifth and sixth grade children (Doctoral dissertation, Fordham University.) *DAI, 34,* 179A.

EMIG–KING WRITING ATTITUDE SCALE FOR STUDENTS. Refer to King, B. (1979). Two modes of analyzing teacher and student attitudes toward writing: the Emig Attitude Scale and the King Construct Scale. Unpublished doctoral dissertation, Rutgers University.

GARY–BROWN WRITING OPINIONNAIRE FOR COLLEGE INSTRUCTORS. Refer to Gary, M., and Brown, S. (1981). Commission on Composition, National Council of Teachers of English. Composition opinionnaire: The student's right to write. Urbana, Ill.: ERIC Clearinghouse on Reading and Communication Skills.

INTERNAL–EXTERNAL CONTROL SCALE. Refer to Fils, K.A. (1981). Changes in self-concept and locus of control as a result of using structured versus unstructured journal writing. (Doctoral dissertation, U.S. International University). *DAI, 41,* 3950A.

MEASURE OF GOALS AND AIMS OF EDUCATION. Nichols, J.G., Potashnik, M., & Nolen, S.B. (1985). Adolescents' theories of education. *Journal of Educational Psychology, 80,* 683–692.

MEASURES OF A WRITER'S CHOICE. Takala, S., Purves, A.C., & Buckmaster, A. (1982). In Fagan, W.T., Cooper, C.R., & Jensen, J.M. (1985) *Measures for research and evaluation in the English language arts,* 2nd edition. Urbana, Ill.: ERIC Clearinghouse on Reading and Communication Skills, National Institute of Education and National Council of Teachers of English.

MOONEY PROBLEM CHECKLIST. (R.L. Mooney & L.V. Gordon, Cleveland, Ohio: The Psychological Corporation.) Refer to Brand, A.G. (1979). The therapeutic benefits of free or informal writing among selected eighth-grade students (Doctoral dissertation, Rutgers University), *DAI, 40,* 1316A–1317A.

NOWICKI–STRICKLAND 40-ITEM INTERNAL–EXTERNAL LOCUS OF CONTROL SCALE. Refer to Rotter, J.B. (1966). Generalized expectancies for internal versus external locus of control of reinforcement. *Psychological Monographs, 80,* 1–28.

QUESTIONNAIRE FOR IDENTIFYING WRITER'S BLOCK. Refer to Rose, M. (1984). *Writer's block: The cognitive dimension.* Carbondale, Ill.: Southern Illinois Univ. Press. ERIC Document Reproduction Service No. ED 236 562.

ROSENBERG SELF-ESTEEM TEST. Refer to Bechtel, J.A. (1978). The composing processes of six male college freshman enrolled in technical programs. Unpublished doctoral dissertation, University of Cincinnati.

SCALE TO MEASURE ATTITUDE TOWARD ANY SCHOOL SUBJECT. (H.H. Remmers.) Refer to Dreussi, R.M. (1976). A study of the effects of expressive writing on student attitudes and exposition (Doctoral dissertation, University of Texas, Austin). *DAI, 37,* 2806A.

SELF-DISCLOSURE SCALES AND SELF-DISCLOSURE AVOIDANCE QUESTIONNAIRE. Refer to Rosenfeld, L.B. (1979). Self-disclosure avoidance: Why I am afraid to tell you who I am. *Communication Monographs, 46,* 63–74.

SEMANTIC DIFFERENTIAL. Osgood, C., Suci, G., & Tannenbaum, P. (1969). Measurement of meaning. In J. Snider & C. Osgood (Eds.). *Semantic differential technique: A sourcebook.* Chicago: Aldine.

TEACHER ATTITUDES TOWARD COMPOSITION INSTRUCTION. Schuessler, B., Gere, A., & Abbott, R. (1981). The development of scales measuring teacher attitudes toward instruction in written composition: A preliminary investigation. *RTE, 15,* 55–63. ERIC Document Reproduction Service No. ED 199 717.

TENNESSEE SELF-CONCEPT SCALE. (W.H. Fittes, Los Angeles: Western Psychological Services.) Refer to Brand, A.G. (1979). The therapeutic benefits of free or informal writing among selected eighth-grade students (Doctoral dissertation, Rutgers University). *DAI, 40,* 1316A–1317A.

TOVATT–MILLER COMPOSITION AND LITERATURE INVENTORY. Refer to Marsh, H.U. (1979). A task-oriented learning group approach to teaching descriptive-narrative-expository writing to eleventh-grade students (Doctoral dissertation, Ball State University, 1976). *DAI, 36,* 7259A–7260A.

WHITELY THOUGHTS ABOUT MYSELF AND SCHOOL QUESTIONNAIRE. Refer to Collins, C. (1978). The effect of expressive writing upon freshmen's cognitive and affective development. ERIC Document Reproduction Service No. ED 173 801.

WRITING APPREHENSION TEST. Daly, J., & Miller, M. (1975). The empirical development of an instrument to measure writing apprehension. *RTE, 9,* 242–249.

2. Audience

CHANDLER ROLE-TAKING TEST. Refer to Rubin, D., et al. (1984). Social cognitive ability as a predictor of the quality of fourth-graders' written narratives. In R. Beach & L.S. Bridwell (Eds.). *New directions in composition research.* New York: Guilford Press, pp. 297–307.

DYMOND EMPATHY TEST. Refer to Rubin, D., et al. (1984). Social cognitive ability as a predictor of the quality of fourth-graders' written narratives. In R. Beach and L.S. Bridwell (Eds.). *New directions in composition research.* New York: Guilford Press, pp. 297–307.

FEFFER ROLE-TAKING TEST. Refer to Rubin, D., et al. (1984). Social cognitive ability as a predictor of the quality of fourth-graders' written narratives. In R. Beach and L.S. Bridwell (Eds.). *New directions in composition research.* New York: Guilford Press, pp. 297–307.

GLUCKSBERG–KRAUSS REFERENTIAL COMMUNICATION GAME. Refer to Kroll, B. (1978). Cognitive egocentrism and the problem of audience awareness in written discourse. *RTE, 12,* 269–281. See also Glucksberg, S., & Krauss, R.M. (1967). What do people say after they have learned how to talk? Studies of the development of referential communication. *Merrill-Palmer Quarterly, 13,* 309–316.

HOWIE–DAY'S METAPERSUASION TASK. Refer to Rubin, D., & Piche, G. (1979). Development in syntactic and strategic aspects of audience adaptation skills in written persuasive communication. *RTE, 13,* 293–316.

3. Cognitive and Inquiry Skills

ADJECTIVE CHECKLIST MANUAL. Gough, H.G., & Heilbrun, A.B. (1965). Palo Alto, Calif.: Consulting Psychologists Press.

A Tentative Criterion-Referenced Test to Measure Thinking Processes. Form A and B (ages 7–11). Refer to Oliver, J.E. (1978). The effects of training in synthesis-level conceptualization upon low-achieving first and fifth graders involved in a literature-based cross-age tutoring program. Unpublished doctoral dissertation, University of Georgia, Athens.

Conceptual Level Test. Refer to Rabianski, N. (1977). An exploratory study of individual differences in the use of free-writing and the tagmemic heuristic procedure. Unpublished doctoral dissertation, State University of New York, Buffalo.

Embedded Figures Test. (H.A. Witkin, 1977. Palo Alto, Calif.: Consulting Psychologists Press.) Refer to Witkin, H.A. (1976). Cognitive style in academic performance and teacher–student relations. In E. Messick (Ed.). *Individuality in learning*. San Francisco: Jossey-Boss.

Insight Scale and Questionnaire. Murray, M. (1987). *Measuring insight in student writing*. Doctoral dissertation, Purdue University.

Iowa Tests of Educational Ability. 7th ed. Feldt, L.S., Forsyth, R.A., & Lindquist, E.F., (1981). Chicago: Science Research Associates.

Matching Familiar Figures Test. Refer to Fischer, O.H. (1981). The relationship between the reflection–impulsivity dimension of cognitive style and selected qualitative and syntactic aspects of time-bound, first draft, expository compositions. *DAI, 41*, 3923A–3924A.

Measure of Topic-Specific Knowledge. Langer, J. (1984). The effects of available information on responses to school writing tasks. *RTE, 18*, 27–44.

Measure of Topic-Specific Knowledge. Newell, G.E. & Mac Adam, P. (1987). Examining the source of writing problems: An instrument for measuring writers' topic-specific knowledge. *Written Communication, 9*, 156–174.

Otis–Lennon School Ability Test. Otis, A.S., & Lennon, R.T. (1982). Cleveland, Ohio: The Psychological Corporation.

Performative Assessments of Writing Tasks: Classification, inductive argument, deductive argument, hypothesis construction, audience adaptation. Faigley, L., Cherry, R., Jolliffe, D., & Skinner, A. (1985). *Assessing writers' knowledge and processes of composing*. Norwood, N.J.: Ablex Publishing Company.

Pertinent Questions. Form A. Berger, R.M., & Guilford, J.P. (1962). Orange, Calif.: Sheridan Psychological Services.

Plot Titles. Berger, R.M., & Guilford, J.P. (1962). Orange, Calif.: Sheridan Psychological Services.

Role Category Questionnaire and Role Construct Repertory Tests. Refer to O'Keefe, B.J., & Delia, J.G. (1976). Construct comprehensiveness and cognitive complexity as predictors of the number and strategic adaptation of arguments and appeals in a persuasive message. *Communication Monographs, 46*, 231–238.

Test of Cognition. Fryburg, E.L. (1972). ERIC Document Reproduction Service No. ED 091 766.

Things Category Test. Catell, R.B., & Taylor, C.W. (1962). Princeton, N.J.: Educational Testing Service.

Torrance Tests of Creative Thinking: Norms-technical manual. (verbal tests, forms A and B and figural tests, forms A and B.) Torrance, E.P. (1974). Lexington, Mass.: Personnel Press.

Verbal Reasoning Subtest of the Dat. Refer to Bechtel, J.A. (1978). The composing processes of six male college freshmen enrolled in technical programs. Doctoral dissertation, University of Cincinnati.

4. Developmental Measures

DEVELOPMENTAL WRITING SCALE. Refer to Graves, D. (1973). Children's writing—research directions and hypotheses based upon an examination of the writing processes of seven-year-old children. Unpublished doctoral dissertation, State University of New York, Buffalo.

DIMENSIONS FOR LOOKING AT CHILDREN'S WRITING AND DRAWINGS IN DAILY JOURNALS OVER TIME. Buxton, A.P. (1976). In Fagan, W.T., Jensen, J.M., & Cooper, C.R. (1985). *Measures for research and evaluation in the English language arts,* Vol. 2. Urbana, Ill.: ERIC Clearinghouse on Reading and Communication Skills, National Institute of Education and National Council of Teachers of English.

THE INTERACTIONAL COMPETENCY CHECKLIST (ICC). Black, J. (1979). Formal and informal means of assessing the communicative competence of kindergarten children. *RTE, 31,* 49–68. See also Black, J. (1979). There's more to language than meets the ear: Implications for evaluation. *Language Arts, 56,* 526–533.

NATIONAL EDUCATIONAL DEVELOPMENT TESTS—subtests of Learning Ability, English Usage, and Word Usage (1970). Chicago: Science Research Associates.

PEABODY PICTURE VOCABULARY. (Dunn, 1959.) Refer to Baden, M.J. (1981). A comparison of composition scores of third-grade children with reading skills, pre-kindergarten verbal ability, self-concept, and sex. (University Microfilms No. 81–22, 588). *DAI, 42,* 1517A.

PIERS–HARRIS CHILDREN'S SELF-CONCEPT SCALE. (1969). Refer to Baden, M.J. (1981). A comparison of composition scores of third-grade children with reading skills, pre-kindergarten verbal ability, self-concept, and sex. (University Microfilms No. 81–22, 588). *DAI, 42,* 1517A.

SCORING GUIDES FOR CHILDREN'S WRITING. Humes, E., et al. (1980). For evaluation of description, narration, exposition, persuasion, personal and business letters, and poems. ERIC Document Reproduction Service Nos. ED 192 371, ED 192 372, ED 192 373, ED 192 374, ED 192 375, ED 192 376.

SEMANTIC FEATURES TEST. Evanechko, P.O. (1970). ERIC Document Reproduction Service No. ED 091 745.

TECHNIQUES FOR COLLECTING LITERACY EVENTS FROM YOUNG CHILDREN. Goodman, Y. (1980). A study of the development of literacy in preschool children. In L. Bird (Ed.). *Occasional papers, program in language and literacy.* Tucson: Univ. of Arizona.

WRITTEN EXPRESSION CHECKLIST. Refer to Baden, M.J. (1981). A comparison of composition scores of third-grade children with reading skills, pre-kindergarten verbal ability, self-concept, and sex. (University Microfilms No. 81–22, 588). *DAI, 42,* 1517A.

5. English As a Second Language Measures

MICHIGAN TEST OF ENGLISH LANGUAGE PROFICIENCY. Division of Testing and Certification, English Language Institute (1962). Ann Arbor: University of Michigan.

STANDARD ENGLISH AS A SECOND LANGUAGE (S-ESL) SPOKEN, GRAMMAR, AND VOCABULARY TESTS. Pederson, E. (1978). New instruments for assessing language proficiency. *Intermountain Teachers of English to Speakers of Other Languages (ITESOL) Papers,* 1981, 2, 18–25. ERIC Document Reproduction Service No. ED 236 648.

6. Essay Assessments

The following assessments are grouped according to type of scoring.

a. Analytic Scoring

ANALYTIC SCALE. Diederich, P.B. (1974). *Measuring growth in English*. Urbana, Ill.: National Council of Teachers of English.

MODELS FOR ANALYSIS OF WRITING. Wilkinson, A., Barnsley, G., Hanna, P., and Swan, M. (1982). *Assessing language development*. New York: Oxford Univ. Press.

SCHROEDER COMPOSITION SCALE FOR ELEMENTARY AND JUNIOR HIGH SCHOOL CHILDREN. Schroeder, T.S. (1973). ERIC Document Reproduction Service No. ED 091 760.

UNIPOLAR SCALE FOR EVALUATING WRITING. Takala, S., Purves, A., & Buckmaster, A. (1982). An international perspective on the evaluation of written composition. *Evaluation in Education, 5*.

b. Holistic Scoring

ADVANCED PLACEMENT ENGLISH TEST. Smith, R. (1975). *Grading the advanced placement English examination*. Princeton, N.J.: Educational Testing Service.

CALIFORNIA STATE UNIVERSITY AND COLLEGES ENGLISH TEST. Long Beach, Calif.: California State University.

ENGLISH COMPOSITION TEST. Conlon, G. (1978). *How the essay in the college board English composition is scored*. Princeton, N.J.: Educational Testing Service. See also Godshalk, F.I., Swineford, F., & Coffman, W.E. (1966). *The measurement of writing ability*. New York, NY: College Entrance Examination Board.

FRESHMAN EQUIVALENCY EXAM. White, E. *Comparison and contrast: The 1974 California State University and Colleges Freshman Equivalency Examination*. California State Colleges Project 75-339, 1973, 1974, 1975, 1976.

c. Primary Trait Scoring

GUIDES FOR EVALUATING EXPRESSIVE WRITING (1977). National Assessment of Educational Progress. ERIC Document Reproduction Service No. ED 205 583.

GUIDES FOR EVALUATING PERSUASIVE WRITING (1977). National Assessment of Educational Progress. ERIC Document Reproduction Service No. ED 205 583.

NATIONAL ASSESSMENT OF EDUCATIONAL PROGRESS. Writing reports No. 3 (1970), No. 5 (1971), No. 8 (1972), No. 11 (1973), 05-W-01 (1975), and 05-W-02 (1976).

NATIONAL ASSESSMENT OF EDUCATIONAL PROGRESS. (1980). The third assessment of writing: Released exercise set. Report No. 10-W-25, 1978–79. Denver, Colo.: National Assessment of Educational Progress, Education Commission of the States.

PRIMARY TRAIT SCORING. Lloyd-Jones, Richard. (1977). In C.R. Cooper & L. Odell (Eds.). *Evaluating writing: Describing, measuring, judging*. Urbana, Ill.: National Council of Teachers of English.

USING THE PRIMARY TRAIT SYSTEM FOR THE EVALUATION OF WRITING. Mullis, I.V.S. (1980). Report No. 10-W-51. Denver, Colo.: National Assessment of Educational Progress, Education Commission of the States.

7. Grammar, Usage, and Syntax

CLOZE TESTS FOR DELETION PRODUCED STRUCTURES. Cosens, G.V. (1972). ERIC Document Reproduction Service No. ED 091 767.

CLOZE TEST TO MEASURE FLEXIBILITY IN WRITING STYLE. Pollard-Gott, L., & Frase, L.T. (1985). Flexibility in writing style: A new discourse level cloze test. *Written Communication, 2*, 107–127.

DEEP STRUCTURE RECOVERY TEST. Simons, H.D. (1969). ERIC Document Reproduction Service No. ED 091 727.

GOLUB'S SYNTACTIC DENSITY SCORE. Golub, L.S. (1973). ERIC Document Reproduction Service No. ED 091 741.

GUIDELINES FOR CATEGORIZING MECHANICS AND GRAMMATICAL ERRORS. (1977). National Assessment of Educational Progress Consultants and Staff, pp. 28–40.

INDEXES OF SYNTACTIC MATURITY. Dixon, E. (1970). ERIC Document Reproduction Service No. ED 091 748.

K-RATIO (KERNAL STRUCTURE INDEX) Calvert, K.H. (1971). An investigation of relationships between the syntactic maturity of oral language and reading comprehension scores (doctoral dissertation, University of Alabama). ERIC Document Reproduction Service No. ED 091 722.

LINGUISTIC CAPACITY INDEX. Brengelman, F.H., & Manning, J.C. (1964). ERIC Document Reproduction Service No. ED 091 753.

LINGUISTIC STRUCTURES REPETITION TEST. Fisher, C.J. (1972). ERIC Document Reproduction Service No. ED 091 746.

LOCUS OF COMPLEXITY IN WRITTEN LANGUAGE. Cayer, R.L., & Sacks, R.K. (1979). Oral and written discourse of basic writers: Similarities and differences. *RTE, 13*, 121–128.

MORPHEME KNOWLEDGE TEST. Shepherd, J.F. (1973). ERIC Document Reproduction Service No. ED 991 730.

RECOGNITION OF LINGUISTIC STRUCTURES TEST. DeLancey, R.W. (1962). Awareness of form class as a factor in reading comprehension (doctoral dissertation, Syracuse University, Syracuse, N.Y.). ERIC Document Reproduction Service No. ED 091 755.

SENTENCE COMBINING SCORING GUIDES. (1977). NAEP Consultants and staff. Refer to Mellon, J. (1981). *Sentence combining skills: Results of the sentence-combining exercises in the 1978–79 national writing assessment*. ERIC Document Reproduction Service No. ED 205 583. Denver, Colo: National Assessment of Educational Progress.

SENTENCE INTERPRETATION TEST. Little, P.S. (1972). An investigation into the relationship between structural ambiguity and reading comprehension. Masters thesis, University of Alberta, Edmonton, Canada. ERIC Document Reproduction Service No. ED 091 715.

SEQUENTIAL TESTS OF EDUCATIONAL PROGRESS. Writing Forms 1A and 1B and Manual (1957). Princeton, N.J.: Educational Testing Service.

SYNTACTIC COMPLEXITY FORMULA. Botel, M., & Granowsky, A. (1972). A formula for measuring syntactic complexity: A directional effort. ERIC Document Reproduction Service No. ED 091 749.

SYNTACTIC DENSITY SCORE. Nold, E.W., & Freedman, S. (1977). *RTE, 11*, 164–174.

SYNTACTIC MATURITY TEST FOR NARRATIVE WRITING. Dauterman, F.P. (1969). Syntactic structures employed in samples of narrative writing by secondary school students. Doctoral dissertation, Ohio State University, Columbus. ERIC Document Reproduction Service No. ED 091 757.

SYNTACTIC MATURITY TESTS. (Fisher, K.D., 1973; Hunt, K.W., 1977) Refer to Straw, S.B., & Schreiner, R. (1982). The effect of sentence manipulation on subsequent measures of reading and listening comprehension. *Reading Research Quarterly, 17*, 339–352.

T-UNIT ANALYSIS. Hunt, K.W. (1965). *Grammatical structures written at three grade levels*. NCTE Research Report No. 3. Urbana, Ill.: National Council of Teachers of English.

Test of Language Judgment. Mantell, A. (1973). ERIC Document Reproduction Service No. ED 991 730.

Type-Token Lexical Density Assessment. Refer to Pederson, E.L. (1978). Improving syntactic and semantic fluency in the writing of language arts students *DAI, 38,* 5892A.

8. Measures for Discourse Types

Delia and Clark Measurement Scales for Cognitive Complexity and Level of Persuasive Strategy. Delia, J.G., Kline, S.L., & Burleson, B.R. (1979). The development of persuasive communication strategies in kindergarteners through twelfth-graders. *Communication Monographs, 46,* 241–256.

Dichotomous Scale for Evaluating Expository Writing. Cohen, A.M. (1973). Assessing college students' ability to write compositions. *RTE, 7,* 356–371.

Dramatic Writing Scales. Brannigan, M., et al. (1977). Holistic evaluation of writing. In C. Cooper, & L. Odell (Eds.). *Evaluating writing: Describing, measuring, judging.* Urbana, Ill.: National Council of Teachers of English.

Evaluation Scale for Personal Information. Refer to Brand, A.G. (1979). The therapeutic benefits of free or informal writing among selected eighth-grade students (doctoral dissertation, Rutgers University). *DAI, 40,* 1316A–1317A.

Explanation Test. Refer to Bass, J.E., & Maddux, C.P. (1982). Scientific explanations and Piagetian operation levels. *Journal of Research in Science Teaching, 19,* 523–541.

Glazer Narrative Composition Scale. Glazer, J. (1971). The development of the Glazer Narrative Composition Scale. Doctoral dissertation, Ohio State University, Columbus. ERIC Document Reproduction Service No. ED 091 763.

Guides for Evaluating Expressive Writing (1977). National Assessment of Educational Progress. ERIC Document Reproduction Service No. ED 205 583.

Guides for Evaluating Persuasive Writing. (1977). National Assessment of Educational Progress. ERIC Document Reproduction Service No. ED 205 583.

Informative Writing Scale. Refer to D'Angelo, J.L. (1978). Predicting reading achievement in a senior high school from intelligence, listening, and informative writing. *DAI, 38,* 2027A.

Literary Rating Scale. Tway, E. (1970). A study of the feasibility of training teachers to use the Literary Rating Scale in evaluating children's fiction writing. Doctoral dissertation, Syracuse University, Syracuse, N.Y. ERIC Document Reproduction Service No. ED 091 726.

Method to Evaluate Expository and Creative Writing (1973). Refer to Brooks, G.M. (1974). An investigation of the relationship of syntactic complexity to reading comprehension. (University Microfilms No. 75-10, 582). *DAI, 35,* 7163A–7164A.

Personal Narrative Writing Scales. Anderson, G., et al. (1977). Holistic evaluation of writing. In C. Cooper & L. Odell (Eds.). *Evaluating writing: Describing, measuring, judging.* Urbana, Ill.: National Council of Teachers of English.

Persuasive Appeals Scale. Connor, U. & Lauer, J.M. (1985). Understanding persuasive essay writing: Linguistic/rhetorical approach. *Text, 5,* 309–326.

Pianko's Evaluative Criteria for Marking Extended Discourse. Refer to Meskin, L.K. (1983). An investigation into the interrelationship between the poetic and transactional modes of language function: A quasi-experimental study. *DAI, 43,* 2911A.

Sager Writing Scale for Creative Writing. Sager, C. (1972). Improving the quality of written composition through pupil use of rating scale (doctoral dissertation, Boston University). ERIC Document Reproduction Service No. ED 091 723.

SCALE FOR DEFINITIONS. Hillocks, G., Kahn, E.A., & Johannessen, L.R. (1983). Teaching defining strategies as a mode of inquiry: Some effects on student writing. *RTE, 17,* 275–284.

SCALE FOR EVALUATING EXPOSITORY WRITING. Refer to Quellmalz, E., Capell, F., & Chou, C. (1982). Effects of discourse and response mode on the measurement of writing competence. ERIC Document Reproduction Service No. ED 236 653. *Journal of Educational Measurement, 19,* 242–258.

SCORING GUIDES FOR CHILDREN'S WRITING. Humes, A., et al. (1980). Primary and intermediate for description, narration, exposition, persuasion, personal and business letters, poems. ERIC Document Reproduction Service Nos. ED 192 371, ED 192 372, ED 192 373, ED 192 374, ED 192 375, ED 192 376.

SCORING WITH AN INFORMATIVE AIM. Refer to Rabianski, N. (1977). An exploratory study of individual differences in the use of freewriting and the tagmemic heuristic procedure. Doctoral dissertation, State University of New York, Buffalo.

TOULMIN ANALYSIS OF INFORMAL REASONING. Connor, U. & Lauer, J.M. (1988). Cross-cultural variation in persuasive student writing. In A.C. Purves (Ed.). *Contrastive rhetoric. Written Communication Annual,* Vol. 2. San Francisco, Calif.: Sage Publications.

9. Program Evaluation

CRITERION-REFERENCED TEST FOR THE ASSESSMENT OF READING AND WRITING SKILLS OF PROFESSIONAL EDUCATORS. Dupuis, M., & Snyder, S. (1979). ERIC Document Reproduction Service No. ED 236 643.

CURRICULUM REVIEW HANDBOOK. Oklahoma State Department of Education (1981). ERIC Document Reproduction Service No. ED 208 540.

EVALUATING COURSE AND TEACHER EFFECTIVENESS. Witte, S., Daly, J., Faigley, L., & Koch, W. (1981). ERIC Document Reproduction Service No. ED 211 981.

EVALUATING INFORMATION IN COMPOSITION. Committee on Teaching and Its Evaluation in Composition of the Conference on College Composition and Communication (1982). ERIC Document Reproduction Service No. ED 236 634.

WRITING CENTER TUTORIAL FORM. Reigstad, R., Matsuhashi, A., & Luban, N. (1980). ERIC Document Reproduction Service No. ED 236 631.

10. Revision Measures

TAXONOMY OF REVISION CHANGES. Faigley, L., & Witte, S. (1981). Analyzing revision. *CCC, 32,* 400–407.

11. Text Analysis

ANALYZING COHESIVE TIES. Hartnett, C. (1980). *Measuring cohesive ties.* Texas City, Tex.: College of the Mainland.

BIPOLAR ADJECTIVE RATING SCALE. Refer to Dreussi, R.M. (1976). A study of the effects of expressive writing on student attitudes and exposition *DAI, 37,* 2806A.

COHERENCE SCALE. Bamberg, B. (1984). Assessing coherence: A reanalysis of essays written for the national assessment of educational progress, 1969–1979. *RTE, 18,* 305–319.

COHERENCE SCALE. Lautamatti, L. (1978). Observations on the development of the topic in simplified discourse. In V. Kohonen & N.E. Enkvist (Eds.). *Linguistics, cognitive learning, and language teaching*. Åbo, Finland: Afinla. See also Lautamatti, L. (1980). Subject and theme in English discourse. In K. Sajavaara & J. Leittanen (Eds.). *Papers in discourse and contrastive discourse analysis*. Reports from the Department of English, University of Jyvaskyla, Finland.

COHESION SCORING GUIDES. National Assessment of Educational Progress Consultants and Staff (1977). Refer to Mullis, I.V.S., & Mellon, J. (1980). Guidelines for three ways of evaluating writing: Syntax, cohesion, and mechanics. Report No. 10-W-50. Denver, Colo.: National Assessment of Educational Progress, Education Commission of the States. ERIC Document Reproduction Service No. ED 205 583.

DISCOURSE-ANALYSIS BASED, WRITTEN, MULTIPLE-CHOICE POST-TEST FOR COMPREHENSION ASSESSMENT OF EXPOSITORY PROSE. (Middle school/Junior High). (P. Young, 1980.) Refer to Kintsch, W. (1974). *The representation of meaning in memory*. Hillsdale, N.J.: Lawrence Erlbaum; and Young, P. (1980). Effect of inference-making aids on poor readers' comprehension. Technical Report No. 555. Madison, Wisc.: Wisconsin Research and Development Center for Individualized Schooling.

DISCOURSE ANALYSIS SCALE. Lautamatti, L. (1978). Observations on the development of the topic in simplified discourse. In V. Kohonen & N.E. Enkvist (Eds.). *Text linguistics in cognitive learning and language teaching*. Soumen sovelletun kielitieteen yhdistyksen julkaisuja, No. 22. Turku, Finland: Afinla.

ESSAY STRUCTURE MEASUREMENT. Refer to Cosda, P., & Maskill, R. (1983). Structure and process in pupil's essays: A graphical analysis of the organisation of extended prose. *British Journal of Educational Psychology, 53*, 100–106.

INSTRUMENT FOR EXAMINING SUBJECT MATTER STRUCTURE. Shavelson, R.J., & Geeslin, W.E. (1975). A method for examining subject-matter structure in instructional material. *Journal of Structural Learning, 4*, 199–218.

MEASUREMENT FOR STRUCTURAL ANALYSIS OF CHILDREN'S COMPOSITIONS. Refer to Stahl, A. (1974). Structural analysis of children's compositions. *RTE, 8*, 184–205.

TESTS OF ANAPHORIC REFERENCE. Multiple Choice Format (TAR-MC) and Cloze Format (TAR-C). Miller, L.A. (1974). An investigation into the relationship of anaphoric reference and reading achievement of grade two children. Doctoral dissertation, University of Alberta, Edmonton, Canada. ERIC Document Reproduction Service No. ED 091 768.

The Use of Human
Subjects in Research

Federal laws

Federal laws exist that provide legal rights to all subjects in research studies of any kind done by institutions that receive federal funds. Composition research is thus subject to these restrictions. The most familiar law, *Family Educational Rights and Privacy Act of 1974*, prohibits any disclosure of educational records or other personally identifiable information without the student's prior written, informed consent. Informed consent means that the researcher must give the subjects enough information about the extent of the public disclosure to enable them to make a reasonable judgment about their concurrence. No coercion can be used to obtain this consent. A researcher cannot make public grades, test scores, pictures, or written products with identifying information unless prior written informed consent has been given. Parents or guardians must give consent for students under the age of eighteen.

Extreme care must be taken to safeguard any research data that can be identified by person. The principal investigator and only a few chosen assistants, briefed about the law, should have access to personally identifiable research data. One of the best ways to achieve this restriction while making the data available for interpretation by other researchers is to list on a main coding sheet two sets of columns: one set that lists the subjects' names and subject numbers, and a second set that lists the subjects' numbers and the data. The first set of columns can then be cut off from the other. If the researcher needs to add missing or additional data, the two sets of coded numbers can be matched. All use of research data must be limited by these rules.

A number of additional principles restrict the collection of data. No undue influence can be exerted on subjects to gain their participation. (Prisoners, for instance, cannot be used.) The subjects must be allowed to withdraw from the research at any time without penalty. There can be no conflict of interest between instructors' professional duties as teachers and their needs as researchers. Disclosure must be made in advance of any potential pain, physical or psychological, associated with the treatments given in the research or the data collection. Prior written consent is needed for research that entails ingestion of materials or breaking of the cutaneous layer, typically not a part of composition research.

An independent committee of researchers, usually a departmental or institutional review board, must review all research proposals that entail ingesting material, breaking the cutaneous layer, or posing potential physical or psychological damage, pain, or discomfort. This committee must decide whether the potential knowledge gains are worth the potential risks. Any research that involves more than minimal risk and is of no direct benefit to subjects must have the potential of advancing knowledge. (Note that this requires a review of prior research on the part of the investigator.) Finally, although this is unlikely in composition research, any mechanical or electrical instrumentation must be designed to be "fail-safe," that is, if the equipment breaks down, no harm will ensue to the subjects.

Instructional exceptions

If the research involves variations on instructional methods designed to achieve the aims of educational courses taught in recognized educational institutions, then prior informed consent may not be needed. This principle exempts a great deal of composition research. But it is wise to have colleagues review one's research design to avoid later conflict-of-interest charges by students. If the goals of the treatment methods in research are reasonable variations of regular instructional methods, then the research functions as a systematic evaluation of the outcomes of instruction.

Other practices exempt from formal review are the use of data from educational tests if the subjects cannot be identified with the data. Survey questionnaires, and interviews are generally exempt from review procedures except where subjects' identification can place them at risk legally or financially. (It should be noted that researchers should not ask questions to which they do not want answers—such as involvement with illegal drugs, illegal sexual behaviors, or other criminal behavior.) Review is also unnecessary for observations of public behavior or the study of existing public data and documents, except where subjects can be identified and placed at risk.

Finally, researchers should keep a full documentation of all actions, records, informed-consent agreements, and so forth for at least three years after the end of the research.

References

Adair, J.G., Dushenko, T.W., & Lindsay, R.C.L. (1985). Ethical regulations and their impact on research practice. *American Psychologist, 1,* 59–72.

Human subjects (March 8, 1983). *Federal Register, 48*(46), 9818–9820.

Student rights in research, experimental activities, and testing (September 6, 1984). *Federal Register, 49*(174), 35318–35322.

Glossary

A weights, in regression equations for prediction research, function as constants of adjustment to accommodate the differences in measurement systems between the predictor and the criterion variables.

Accuracy. See Precision.

Alpha, Cronbach's alpha is a correlational index of internal consistency reliability for use with interval and nominal variables.

Alternative hypotheses pose alternative explanations that could explain or account for the phenomena under study.

Analysis of variance, ANOVA, a method of statistical analysis broadly applicable to a number of research designs, is used to determine differences among the means of two or more groups on a criterion variable. The condition and treatment (independent) variables are usually nominal (or made into nominal) variables, and the criterion variable is an interval variable. If there is more than one criterion variable, then the analysis should be a multivariate analysis of variance.

Analytic evaluation gives a separate score for each of several qualities of a piece of writing.

Attitude scales often use the Semantic Differential or a Likert-type scale to assess feelings about writing or its pedagogy.

B weights are regression weights used when variables are in raw-score form, in other words, have not been converted into standard score variables with means of zero and standard deviations of 1.00. In a regression equation, each raw-score variable (grades, essay ratings, or GPAs) gets an appropriate B weight according to its standard deviation and its correlation with the criterion and other variables in the equation.

Beta weights in a regression equation are the regression weights used for predictor and criterion variables when they are stated in standard score form, for example, all with means set at zero and standard deviations set at 1.00.

Case studies are qualitative, descriptive studies that examine one or a few subjects using a variety of data collection methods: interviews, records, tapes, collections of papers, and so forth. Their major purpose is to identify important variables.

Chi Square (χ^2), a statistical analysis method used to determine the relationship between two or more nominal variables or the differences between two or more groups of subjects on one nominal variable used as a criterion variable.

Classification is a descriptive research design used to forecast a person's compatibility with a group. Discriminant analysis (see below) is used.

Cluster sampling uses observational units of whole classes rather than individual students.

Coding, or content analysis, is the process of interpreting data by identifying and operationally defining constructs, which become the variables of the study.

A **concept** is the lay term for a construct, an abstract aspect of behavior. See Construct.

Confidence limits are the possible range of percentages of average incidence of a variable (plus-and-minus percentages around a sample mean in a population) based on evidence from a sample. See Population; Sample.

A **construct,** the scientific term for a concept, is the abstract label for an aspect of behavior. It is an inductive summary composed of many features, that postulates how the aspect will operate in certain situations, how it is distinguished from other constructs, and how it can be measured.

A **control group** in experimental research is the group to which the experimental treatment is not applied. It functions as the standard for comparison on the criterion variables.

Correlation is a common statistical analysis, usually abbreviated as **r,** that measures the degree of relationship between pairs of interval variables in one sample. The range of correlation is from -1.00 to zero to $+1.00$.

A **correlational matrix** shows the correlations between all pairs of **K** variables across the subjects in a sample.

Covariance, ANCOVA, is a product of the correlation of two related variables times each of their standard deviations. In statistical analyses of true experiments, covariance is used to reduce the variance of a criterion variable by means of its correlation with a previously measured control variable, thereby enhancing the capacity of the researchers to detect differences between treatment effects in a criterion variable.

Criterion variables for experiments are the dependent variables used to detect differences between groups after the treatment has been administered.

Criterion variables for prediction are the quantified indicators of success or achievement in a course of instruction. They measure the behavior for which predictor variables are sought.

A **data matrix** is a vertically oriented rectangular array of data that organizes scores for each individual for all variables under study. On the left (vertical) side of the matrix, "n" represents the number of subjects. Across the top, "**K**" represents the variables of all types in the study. For prediction and classification studies, the **K** variables are divided into two sets: predictor and criterion variables.

Degrees of freedom are the basis on which statisticians enter statistical tables. They are usually determined on the basis of the number of observations in a set of behavior or the number of categories in which the data have been placed. (See Appendix I.)

Dependent variables are those used as criterion measures for experiments and descriptive research.

Discriminant analysis is a statistical method of classifying subjects into similar groups as categories within nominal variables.

Effect size is the difference between the mean of an experimental and control group criterion variable (in true and quasi experiments), divided by the standard deviation of the variable. Effect size is a standard z score with a scale from about -3.00 to zero to $+3.00$.

Ethnographic research examines an entire environment, looking at subjects in their natural context and determining the relationship among variables and their effect on subjects.

External validity refers to the generalizability of research results across subjects, treatments, conditions, and criterion variables.

F-Test is a statistic used primarily in analysis of variance to report the relative equality of a set of means. The larger the **F** value, the more unequal are the set of means.

Formative evaluation takes place while a program is in operation, sometimes even in stages of development. Its purpose is to identify strengths and weaknesses while the program is being conducted so that immediate improvements can be made.

Frequency counts are the number of items, data, and subjects that belong in any category of a nominal variable.

Holistic evaluation assigns a single rating to a piece of writing, either grouping it with other graded pieces or scoring it on the basis of a set scale.

Homogeneity of treatment variables is their similarity in effect on criterion variables, particularly in meta-analyses. Homogeneity of variance is an assumption in analysis of variance that all variances within treatment condition cells are essentially equal.

Hypotheses are tentative explanations based on theory to predict relationships, often causal, between variables.

Independent variables are those that are part of the research situation: the treatment, the experimental conditions, the characteristics of the subjects (age, level of development, sex, etc.).

Interactions in the analysis of variance are the combined effects of two or more treatment or condition variables. A significant interaction says that something is happening synergistically between or among levels of two or more variables, causing either greater or lesser effects in combination than would be expected from the simple summative effect of the variables separately.

Internal validity is the ability of the researcher to make cause-and-effect statements about criteria differences among groups as the result of treatments or conditions, particularly in true and quasi-experiments.

Interval data are variable score categories assumed to be equal-appearing intervals psychologically. They typically are assumed in such variables as test scores, ratings, and grades.

Interval measurement is quantification of data in which the score categories are expected to be equally distant psychologically from one another. Most grade scales (A, B, C, D, F) and holistic scales (1–5), if developed with care, can be considered equal-interval scales.

Kappa (Cohen's) is an accurate index of agreement among judges after chance influences have been removed. It ranges from zero to $+1.00$.

KR 20 (Kuder–Richardson 20), a correlational index of internal-consistency reliability, is used only with nominal variables (enumeration data). **See** Cronbach's alpha (Appendix I) for the more general index of internal-consistency reliability.

The **mean** is the sum of all scores for a variable, divided by the number of subjects—in lay terms, an average.

Measurement is the process of quantifying variables. Standardized measures often have normative standards, reasonably known reliabilities, and some evidence of validity. Some frequently used direct-measurement instruments for writing include: holistic, analytic, and primary trait scoring of essays and T-unit analysis. Indirect measures include various standardized tests of grammar, punctuation, style, or usage.

Meta-analysis is a systematic, replicable, and relatively unbiased method of summarizing the overall results of a particular body of experimental research literature. It uses effect sizes to assess the magnitude of the results of several studies and can enhance the generalizability of research conclusions.

A **Model** is a specification of the theoretical components of a behavior such as writing. This usually includes identifying the major variables and how they interrelate. Three general types of models are linear, process, and simulation models.

Multiple correlation is the statistical interrelating of a set of predictor variables to predict maximally a criterion variable. The prediction method resulting is a regression equation, which optimally weights (with B's or betas) each predictor variable and adds any constant, A, that may be necessary.

Nominal data are those that result from simple counting or enumeration by nominal categories of data. These counts are often then transformed into percentages or proportions.

Nominal measurement is simply counting frequencies, by classifications (types of writing: essays, poems, novels, biography). These frequencies are readily transformed into proportions and percentages.

Nonparametric statistics are analytical methods that generally do not make assumptions about the form of the distribution of the data, such as normal distribution.

A **null hypothesis** is a formal statistical statement that there is no difference between groups on a variable or no relationship among variables in a group of subjects. In classical statistical analysis, researchers reject or accept this null hypothesis via the statistical analysis of their data.

One and two-tailed statistical tests are used with tests like the t-test to predict whether one group's population means are larger or smaller than another population's (the one-tailed test) rather than simply different (the two-tailed test).

An **operational definition** specifies how a variable will be observed, produced, or measured.

An **ordinal scale** places people, objects, or concepts in an ordered relationship to one another on one dimension or variable. These ranks should not necessarily be considered equally distant from one another.

Parametric statistics are analytic methods based on defined distribution of observations, usually a normal distribution.

A **participant observer** is a researcher who, with a minimum of overt intervention, becomes a member of the classroom or natural environment being studied.

A **percentage** is a frequency count in a category divided by the frequency counts in all the categories and multiplied by 100.

Percentage of agreement is an index of agreement among two or more raters or coders. It is a simple percentage of the number of times raters agree on a rating divided by the total number of items rated.

Percentile scores are rank-ordered scores based on the frequency of scores in a measurement system in a sample of subjects. Percentile scores range from 0 to 100 percent.

Phi coefficient is a statistical product-moment correlation of two dichotomous nominal variables like gender (male/female), each of which are treated as interval variables by labeling one of their two categories as "1" and the other as "2."

Point biserial correlation is a statistical product-moment correlation of two variables, one a dichotomously measured nominal variable and one an intervally measured variable.

The **population** is the total number of subjects or objects or treatments to which researchers want to generalize their sample results.

The **precision** of the conclusions about a population from which a sample has been drawn in a sampling survey is the "tightness" of the confidence limits. (**See** Confidence Limits.) The larger the sample, the smaller the range of the confidence limits.

Prediction research uses a set of predictor variables to forecast subjects' future performance on an interval criterion variable. Regression and correlation methods are used.

Predictor variables in prediction research are used to forecast future performance. Usually they are *interval* variables such as prior GPA or grades in prior English courses or dichotomous nominal variables such as male/female.

A **pre-experiment** is the application of a treatment to a single group and the use of pre-tests and posttests, but with no control group for purposes of comparison. Little cause-and-effect knowledge can be gained from this exploratory research.

Primary trait scoring gives separate scores on criteria that have been established by analyzing the assignment and the resources used by writers in fulfilling it.

Program evaluation is an assessment usually undertaken for administrative or instructional design purposes to determine such issues as the degree to which a writing program is achieving its goals, is more effective than an alternative curriculum, is efficiently run, is academically sound, and so forth.

A **proportion** is the frequency count in a nominal category divided by the total frequencies in all categories observed. A proportion is always a decimal number ranging between zero and 1.00. The sum of all proportions for all categories should be 1.00.

Protocol analysis is an analysis of a transcription made from a subject's oral description (thinking aloud) of activities, ordered in time, in which a subject engages while performing a task.

A **pseudoexperiment** is a faulty experiment in which no attempt is made to ascertain the initial status (equal or unequal) of treatment and control groups.

Qualitative descriptive research closely studies individuals, small groups, or whole environments in order to identify important aspects or **variables** of any phenomenon to be examined. Qualitative researchers typically neither establish treatment or control groups nor quantify variables.

Quantitative descriptive research isolates systematically the most important variables developed by case studies and ethnographies, further defines them, quantifies them at least roughly, if not with some accuracy, and then interrelates these variables by means of statistical analysis.

Quasi-experiments, much like true experiments, have subjects, treatments, and criteria, but use intact, nonrandomized groups. Researchers apply pretest and posttests and establish hypotheses to declare the possible results of ineffective treatments because of the influence of threats to internal validity.

Quota sampling in survey research is the selection of subjects proportionate to the percentage of incidence of some feature of the population, for example, sex, race, college level. Quota sampling is used where populations are not readily numbered or in some sequence.

Random sampling in survey research is a method of drawing samples strictly by chance and yielding no discernible pattern beyond that expected by chance. Done strictly as directed, random sampling yields samples that are representative, within certain confidence limits, of the population from which they are drawn.

Randomization in experiments is used to allocate subjects to various experimental and control groups. The subjects within the groups can be considered initially "not unequal," because they have been selected by a random process.

The **range** is a statistic defined as the highest score in a sample minus the lowest score (and sometimes plus one).

Rank order measurement simply organizes categories on a single dimension systematically in relationship to one another from lowest to highest, or vice versa, and numbers them in order. The resulting ranks are not necessarily assumed to be equal psychological intervals, although the ranks are often treated statistically as if they were.

Rank order statistics, used with rank order data, show relationships among variables in a group of people or differences between and among groups of people on scores of rank-ordered variables.

Ratio measurements have the properties of interval measurement plus a true zero, a fixed point at which none of the dimensions exists.

Regression equation is the algebraic expression of the correlational relationships among measures that are used to predict a criterion measure's scores. This equation consists of weights (B or beta) for each predictor measure plus a constant of adjustment (A).

Reliability, an index of the precision of measurement of an instrument or of observers, is the ability of independent observers or measures to agree. It is generally reported as a decimal, stated as a positive correlation coefficient that will range from .00 to close to 1.00. There are three types of reliability: equivalency, stability, and internal consistency. A percentage of agreement among raters is also a measure of reliability.

Equivalency reliability is the degree of relationship between the scores on two sets of alternate forms of a test for a group of subjects who take both tests at the same point in time. To be equivalent, their average scores and standard deviations must be the same, as well as their average intercorrelations among items.

Internal-consistency reliability, the most theoretically important concept of reliability, is the degree to which a test measures a single dimension of human ability or achievement. It is also an index of the precision of the measurement instrument or of the observers. This kind of reliability is necessary for the interpretation of most data and especially correlations. Cronbach's alpha is used to determine internal-consistency reliability. (The Kuder–Richardson 20 is a special case of Cronbach's alpha, used with dichotomous data only.)

Interrater reliability is a form of internal-consistency reliability indicating the amount of correlation among raters.

Stability reliability is the correlational relationship of variables over time.

A **sample** is a subset of population.

Sample size is the number of people, compositions, etc. in a sample group, determined by balancing two factors: the degree of imprecision acceptable for decisions and conclusions, and the cost of collecting data and analyzing them.

Sampling studies are descriptive research methods used to describe a large population in terms of a **sample,** a smaller part of that group.

Scatter diagram is a graphic representation of a correlation used to present the scores of a group of subjects on two variables in order to show their visual relationship.

Shrinkage is the term used in prediction research to explain the reduction of predictive capability when researchers shift from the prediction-establishing sample to future samples. The equation developed usually does not predict relationships in future samples quite as well, especially if the ratio of the original number of subjects to the number of variables was not large. For the best prediction, the correlation or multiple correlation should be reduced, for example, "shrunk," by means of a shrinkage equation.

Significance in statistics is a statement of the degree of rarity of a result based on chance probability alone. In almost all cases of statistical analysis, the calculated statistic is compared with standard, tabled values of chance distribution of that statistic. If the calculated value is sufficiently larger (for almost all statistics) than the tabled value, the result is called statistically **significant,** meaning that the statistical relationship between two variables observed is unlikely to have occurred simply by random chance alone. Significance is declared usually at a probability (p) level of five or one percent.

The **Spearman–Brown formula** shows the relationship between the length of the test (number of items) or the number of observers and the internal-consistency reliability, r_{ii}, of the test or measurement methods.

The **standard deviation** is the statistical indication of the measure of individual differences or dispersion of scores for each variable. For very large samples there are about six standard deviations in a range of scores—three up and three down—from the average for a variable.

Standardized tests, often multiple choice, are objectively scored (i.e., reliable and unbiased) tests and have scores from normative samples for interpretation purposes.

Stratified sampling in survey research is the selection of a sample from a stratum or part of a population, which becomes the population in itself.

Strong quasi-experiments are those experimental research types in which the pretests or initial observations show that the groups are equal on the variables measured.

Successive-category measurement is a measurement system that results in ordered categories from ratings, grades, interviews, and questionnaires. If carefully constructed, these categories yield interval measurements.

Summative evaluation assesses the outcomes of an educational program, usually in comparison with other programs, in order to make administrative decisions about the overall value of the program, its continuance, faculty, instructional effectiveness, and levels of funding.

Surveys are methods often used in descriptive research to gather responses from a large group of subjects on a relatively small set of variables of interest.

Systematic sampling, a type of random sampling used when the population is already organized in a sequence, establishes an order in advance to account for all units to be studied and observes every "**nth**" sampling unit (say fifth, or tenth, or twentieth), the starting point of which is chosen at random from within the sampling unit.

t-Test is a statistic used basically to determine whether the means of two groups are equal or unequal on a variable.

A **theory** is a conceptual framework, a set of generalizations drawn from facts (which are statements about interrelationships among variables), devised to explain behavior.

Thick descriptions are detailed, extensive accounts of writing and other behavior and its rich context, used to report the results of ethnographic research and case studies.

Threats to external validity are those circumstances that prevent a researcher from generalizing the cause-and-effect results in a sample, a treatment condition, or a measurement system to other samples, treatment conditions, or measurement systems. Two types of threats are defined here.

Interaction threats arise from combinations of threats to external and internal validity, which singly may have no effect.

Reactivity arrangements are the factors in the experimental situation that are unrepresentative of real-world conditions, such as the artificiality of the experimental setting, the subjects' knowledge that they are participating in an experiment, and attempts by the subjects to outguess the researchers.

Threats to internal validity flaw the cause-and-effect properties of research designs. Some of the major threats are listed here:

Compensatory equalization of treatments occurs when administrators or others demand that the "benefits" of the treatment be extended to the control group.

Compensatory rivalry and **resentful demoralization** occur when subjects have emotional responses to the experimental or control conditions.

Diffusion, at one time called "leakage," occurs when participants of one group communicate with those of another group about the research methods.

The **history** threat occurs when extraordinary surrounding, external conditions cause either the control or experimental group or both to change in ways that they ordinarily would not. Fires, riots, and major social changes are example of this threat.

The **instability** threat is the sum total of many factors in behavioral science research such as the fluctuation of human variables over time, the imprecise measurement systems, and the natural variations in applying experimental and control treatments, which add to the variability of the criterion measures used.

Instrumentation involves changes in the measurement system (on the same dimension of behavior) such as the lack of good calibration of alternative forms of measurement instruments or systematic differences among observers' ratings. All of these can cause biased results.

Maturation is the effect on subjects of experience and everyday growth, both biological and psychological.

Mortality is any reduction (or sometimes increase) in the numbers of the experimental or control group subjects.

Regression-toward-the-Mean, a major threat to the internal validity of the research design, occurs when the researcher interprets the observed score to be farther from the mean than the subject's true score actually is. This is a threat especially for subjects at the extremes of the populations—remedial and honors students—and on less reliable measures.

Selection occurs when a researcher claims that a treatment made a difference among groups that differed before the treatment was given.

Testing is a seeming increase in scores, which is the result of giving subjects the same or similar sets of tests or observations more than once. This can occur when pretests are used also as posttests.

A **treatment** is an instructional strategy, a classroom environment, or some condition applied by the experimenter to a treatment group.

Triangulation is the combining of multiple sources of data such as field notes, videotapes, journals, interviews, and written work to assess a variable.

A **true experiment** is a research design in which the researcher systematically applies a treatment to a group and withholds it from a second group in order to determine what its effect is on the criterion variables. Randomization is used to allocate subjects to treatment and control groups. (Treatments can also be randomly assigned to these randomized groups.)

T-Unit is a main clause plus any subordinate clauses or nonclausal structures attached to it or embedded in it.

T-Unit analysis entails counting the number of T-units (one main clause plus any subordinate clauses attached to it) in a piece of writing and computing the average length of the T-unit by counting the number of words in forty-five T-units selected from three or more samples of a given student.

Type I errors in statistical test interpretation occur when a researcher calls a result on a sample statistically significantly different, which in fact in the population is not so, but is due rather to chance. If researchers want to avoid this type of error, then they should set the p level low, at $p = .01$, where results that are improbable are expected to happen rarely, only one time in one hundred.

Type II errors occur when researchers are unable to detect statistically significant differences between groups or relationships among variables within a group when in fact they truly exist. To avoid this error, researchers can set the p values at larger points than the traditional .05 and .01, for example, at .10 or .20. These would be acceptable p values, especially in introductory research when an investigator has a small sample or less than highly reliable measurements of variables.

The **validity of measurement systems,** tests, observations, interview results, and so forth, is their authenticity, their ability to measure whatever they are intended to assess. There are four types of validity: concurrent, construct, content, and predictive.

Concurrent validity is the degree of relationship between a known, valid measure of behavior and an alternative measurement system.

Construct validity is the measure's congruence with the theoretical concepts of a field, the extent to which the measured variable accurately and adequately accounts for the behavior or characteristic it attempts to describe.

Content validity is the extent to which the topics, items, and questions in a test are representative of the educational or instructional requirements of a curriculum or a course for which the test is being used as a placement tool or a measure of achievement.

Predictive validity is the extent to which a measure forecasts some future condition, score, or judgment of behavior like course grades.

A **variable** is a unified concept, abstract property, or dimension of an object or behavior under study that is measured.

Control variables are those brought under control by being held relatively constant, by having their influence reduced by means of statistical manipulation, by being eliminated entirely, or by having their influence equated across groups.

Dependent or **criterion variables** are those by which the outcomes of a study are judged (e.g., tests, holistic ratings, T-unit counts).

Extraneous variables are those that produce unpredictable influences on the outcomes of research. They make outcomes more difficult to predict, more ambiguous. The threats to internal and external validity are considered extraneous variables, which must be minimized by the research design.

Independent or **treatment variables** are those that are observed, changed, or manipulated systematically by the researcher to see how these changes affect the dependent or criterion variables.

Intervening variables are those that produce an effect indirectly by mediation, that is, a process of actuating hypothetical effects on people.

Manipulated variables are treatments applied in experiments, or research conditions or types of subjects in qualitative research.

Moderator variables are those that influence or moderate the effect or impact of a treatment variable on a criterion variable.

Variance is a statistical concept that represents the degree of individual differences among people on a variable. It is the standard deviation squared.

Weak quasi-experiments are experimental research designs that have initially unequal groups.

z **Scores. See** effect size.

Glossary of Symbols

χ^2 Chi-Square statistical analysis, resulting from frequency (or counted or enumeration) data.

C   The symbol (C) for the contingency coefficient, a measure of the strength of the associations between two nominal variables. Used after the Chi Square analysis.

η A measure of the degree of correlation between two intervally measured variables that does not depend on a linear relationship (eta).

F   The statistic usually resulting from an analysis of variance, of interval criterion data categorized on a nominal variable. Used in honor of Sir Ronald Fisher, who developed the analysis of variance and the concept of true experiments.

H   Often Kruskal–Wallis's H, the symbol of a rank order (considered non-parametric) statistical analysis analogous to the analysis-of-variance F-test.

H   In the meta-analysis developed by Hedges, a measure of the homogeneity of variance on the criterion variable within each of the treatment variable conditions or treatments.

λ A symbol (lambda) of a statistic used in multivariate analysis to indicate strengths of relationship among variables. It is unusual in statistical notation because, as it gets smaller, it tends to be more significant; Wilks' lambda.

M   A symbol indicating the mean.

n The number in a sample.

N   The total number or sum of all the n's.

ϕ Cramer's phi, the measure of association for strength of the relationship between two nominal variables. Cramer's ϕ should now be used instead of C.

p The abbreviation for the index of probability. Probability values run from 0.00 to 1.00, 0.00 being no probability of an event occurring and 1.00 being certainty that an event must occur. The "p" values for statistical significance most commonly used are $p = .05$ and $p = .01$.

Q Cochran's Q, used as the statistic for a correlated-data (repeated-measures) Chi Square analysis.

r The symbol for correlation.

R The symbol for multiple correlation, with more than one predictor variable.

ρ The Greek letter rho, for the Latin letter, "r," used for the symbol of the product-moment correlation between two variables, both quantified in rank order form.

r_b The symbol for biserial correlation, the estimate of the product moment correlation between dichotomously quantified variables and an equal-intervally quantified variable. (Should now seldom be used.)

r_t The symbol for tetrachoric correlation, the estimate of the product moment correlation between two dichotomously quantified variables. (Should now seldom be used.)

σ The small Greek letter, sigma, used as the symbol for the standard deviation.

Σ The capital Greek letter for sigma, used to indicate a sum. $\Sigma\, X_i$ indicates the sum of all the scores in classification "i."

S Another symbol for standard deviation. (Usually the Greek letter stands for population values of the standard deviation and Latin letter for the sample values—but not always.)

σ^2 The symbol for variance.

S^2 Another symbol for variance.

τ The small Greek letter, tau, used as the symbol of a true nonparametric measure of association between two variables, both rank ordered; Kendall's tau.

U Often the Mann–Whitney U, the statistic used with rank order data (considered nonparametric) analogous to the t-test.

ω A measure of the strength of the agreement among judges' rank scores (considered nonparametric); Kendall's ω.

X The letter X symbolically stands for an variable. With a subscript "i," a given score or class of score on that variable.

$\bar{X}$ The mean for the variable X.

z The fundamental metric of psychological variables, having a mean of zero and scores having a standard deviation (and variance) of 1.00.

Name Index

Subject Index